Peking Man Searcher Indicted

CHICAGO — Financier Christopher Janus's widely publicized search for the bones of prehistoric Peking Man was a $640,000 fraud, a federal grand jury charged Wednesday.

News Watch

The grand jury indicted Janus, 69, on 37 counts of bank and mail fraud, accusing him of fraudulently obtaining $526,000 in loans from banks and another $120,000 from investors.

The bones, uncovered by archaeologists in Peking in 1926, vanished from China during World War II. Janus, a Chicago investor and world traveler, contends the Chinese government asked him to direct a search for the bones in 1972, shortly after the United States resumed diplomatic dealings with the communist nation.

The Search for Peking Man

Christopher G. Janus
with William Brashler

Macmillan Publishing Co., Inc.
New York

Macmillan Publishing Co., Inc.
866 Third Avenue, New York, N.Y. 10022
Collier Macmillan Canada, Ltd.

Library of Congress Cataloging in Publication Data

Janus, Christopher George, 1911–
    The search for Peking man.

    Includes index.
    1. Sinanthropus pekinensis. I. Brashler, William, joint author. II. Title.
GN284.7.J36   573'.3   75-4693
ISBN 0-02-558990-3

Second Printing 1975

Printed in the United States of America

To the descendants of Peking man—
all 800 million of them.

Acknowledgment

I am most grateful for the help of William Brashler and Joan Ettlinger, two very gifted writers. Joan took time out from writing her own book on British economist Joan Robinson to edit and transcribe my tapes which form the basis of *The Search for Peking Man*. Joan introduced me to Chicago writer William Brashler who, after considerable research and additional interviews with many of the persons involved in the search, wrote the book.

It is a pleasure to express my appreciation to Fred Honig of Macmillan for his valuable suggestions and for the superb editing he did on the book.

Contents

CONTENTS

Preface

In the following pages, Christopher Janus will tell the tale of an unfinished quest which takes you over one huge ocean and across several countries; a real-life chase as stirring and as exciting as a television thriller. But his quarry is not a villain to be apprehended after a toothpaste commercial at five minutes before the hour. Nor is it a pot of gold, but several sea-chestsful of 400,000-year-old bones, the remains of hunters and berry-pickers who lived rewarding lives when man was bright-eyed and young but life was already old.

These long-fossilized bones are sacred. They form part of the precious heritage of mankind as a whole. They mutely tell us how one branch of our species rallied; they help to show how mankind arose from its simple technology to meet the challenges of today's overcomplicated, computerized world.

And they are uniquely, personally, holy to the Chinese people, wherever they live, for a very special reason: *the Chinese consider them to have been their very own ancestors.*

As a physical anthropologist who has also worked on this problem, I agree with the late, gifted, anatomist Franz

Weidenreich, who studied the original specimens meticulously before they were lost in the tumultuous tragedy of Camp Holcomb: that the Chinese people's belief was well-founded.

Mr. Janus's quest is not just a middle-aged schoolboy lark, nor a simple junket. He is a man of much education and high purpose. His serious, philanthropic enterprise has led him into situations, sometimes dangerous, always stimulating. Don't underestimate him. And, by all means, don't start reading this book too late at night if you are due in the office at nine.

CARLETON S. COON

Gloucester, Mass., 1975

*"It is indeed a good thing to
be well-descended, but the
glory belongs to the ancestors."*

—PLUTARCH (A.D. 46–120)

Prologue

About thirty miles southwest of Peking, in a dusty barren area known as the Petchili Plain, there lies a small ridge of mountains of which one peak is called Dragon Bone Hill. The region is a rich source of animal and human bone fragments, which the Chinese believe have miraculous healing powers. For hundreds of years peasants combed the hills and adjoining plains in search of the bones. Many of the fragments were sold to apothecaries who ground them into powder which they sold as medicine.

In 1899, a young German naturalist, K.A. Haberer, became interested in the "dragon" bones and began roaming the quiet Chinese countryside on the heels of the peasants. In a short time he had amassed more than a hundred different types of bones. Some he found himself; others he purchased from apothecaries. What interested Haberer about the bones had nothing to do with medicine at all, but with anthropology; he suspected the fragments to be fossilized remains of creatures who had inhabited Dragon Bone Hill centuries before.

Most of Haberer's fossil fragments were from hyenas, wild boars, and other animals once common to the region, but one of his specimens stood out from the rest. It was a

13

tooth, not of a gibbon or a tiger, but, he was quite sure, of a human, what he thought to have been an apelike man.

In 1903, Haberer brought his entire collection to a Munich anthropologist who confirmed his belief that the tooth was human. Encouraged by this opinion, Haberer returned to China to trace the origin of the tooth. He finally determined that it was among the fragments he had purchased from an apothecary who, in turn, obtained most of his fossils from peasants from the region of Dragon Bone Hill.

Haberer's specimen was mysteriously lost some months after it had been examined in Munich. It may be overly dramatic to suggest that this loss foreshadowed the later fate of Peking man, but it is an intriguing notion.

It was almost twenty years before the anthropological community responded to Haberer's find. The first man to pick up the trail was a Swede, Gunnar Andersson. Andersson began his investigations near the village of Choukoutien, on the outskirts of Peking, and by 1918 had determined that the nearby Dragon Bone Hill was indeed, as Haberer had suggested, a goldmine of fossil remains.

In 1921, Andersson obtained financial backing from the Swedish Academy for research and excavation at Choukoutien. His assistant, Dr. Otto Zdansky, began digging in the area pinpointed by Haberer and he too uncovered hosts of fossilized animal remains. His finds were sent to Uppsala, Sweden, for laboratory analysis, and among them two specimens, a bicuspid and a third upper molar, were identified as unmistakably human.

With Zdansky's discovery, Andersson was able to draw

the interest of other anthropologists, including those working at the Peking Union Medical College (PUMC), a small but noted institution generously supported by the Rockefeller Foundation. On the strength of Zdansky's new specimens, the Rockefeller Foundation and the Geological Survey of China agreed to establish, equip, and support the Cenozoic Research Laboratory at the Choukoutien site. In 1926, teams of excavators arrived at Dragon Bone Hill and began their search for Haberer's apelike man. They called him *Sinanthropus pekinensis*, Peking man, a creature they believed had roamed the ridges of Dragon Bone Hill as long as 400,000 years ago.

For fifteen years, Peking man was picked from his grave piece by piece: a tooth, a skull, a jawbone. The diggers detonated mild charges, prodded gently with ball-peen hammers, and sifted the hill's crushed limestone grain by grain until more than 175 specimens were unearthed. Each one was analyzed, charted, photographed, and cast in plaster. Researchers estimated that as many as forty individuals were represented by the find.

Work at the Choukoutien site came to a halt, however, in 1937. The rumblings of World War II were heard in the distance and the threat of an invasion by hostile Japanese forces became a reality. It was decided by those in charge of the project that the bones of Peking man should be transported to the United States. Careful preparations were made for the evacuation. The treasure was meticulously packed and consigned to a United States Marine escort.

The plan was that Peking man would wait out the war in the safekeeping of the American Museum of Natural History in New York. Once peace was restored, the research at Choukoutien could continue. That was the plan, but when the smoke of war cleared and much of the Far East was reduced to rubble, Peking man was nowhere to be found.

PART ONE

The Discovery
and the Disappearance

CHAPTER 1
Davidson Black

Davidson Black, a gaunt-faced, deadly serious anatomist from Canada, was to contribute more than any other individual to the Choukoutien digs. In his early forties, Black was considered an expert on anatomy and a leading authority on human skeletal remains. He was a professor at Peking Union Medical College, in whose laboratories much of the fossil research had been conducted, and had collaborated with Andersson on reports of the first Choukoutien finds. But more than his scholarship, Black brought to the digs incredible zeal and devotion; he was the driving force behind a major breakthrough in man's knowledge of his origins.

In 1926, Black and Andersson formally announced that the teeth found by Zdansky might be as many as one million years old. Their claim, based on Black's deductions, was a bold one, for that period of prehistory is thought to be the dawn of mankind, the time when mammals replaced reptiles as the dominant form of animal life. Black was not totally certain of his estimate, and was later to modify it, but he was positive that his specimens provided proof that early man had inhabited eastern Asia.

Black's claim was given worldwide attention. His arguments for the two teeth were analyzed and debated at length not only in scientific journals but in general circulation magazines and newspapers. In the United States, the debate came in the wake of the celebrated "Monkey Trial." In July 1925, a young Tennessee school teacher named John T. Scopes was convicted of breaking state law by teaching Darwin's theory of evolution. In that trial the biblical version of creation won the day but the precepts that led to a guilty verdict were undermined by the findings of Black and Andersson.

Until the discoveries at Choukoutien, anthropologists had only the work of Dutch anatomist Eugene Dubois on which to base theories of the presence of early man in Asia. In 1891, Dubois found a skullcap and, one year later, a thigh bone in the gravel of the Solo River in Java. He named his discovery *Pithecanthropus erectus*, from the Greek meaning "standing ape-man."

Dubois' Java man, as he is popularly known, was thought to be 700,000 years old. He was hailed by some as the missing link between ape and *Homo sapiens*, and dismissed by others as nothing more than a true ape. The scarcity of the find prevented any firm conclusions about the nature of Java man and, because the bones were found in aquatic deposits, it was unlikely that diggers would ever uncover more remains in that location. Java man had been transported from his natural habitat by millennia of shifting tides. The discovery of Peking man in a dry-land grave suggested exciting possibilities; in all likelihood he and others had lived and died on Dragon Bone Hill.

In 1927, a full-scale excavation plan was devised by Black

and Dr. W.H. Wong, director of the Geological Survey of China. The field work was put under the direction of a German expert, Birger Bohlin. Bohlin's team worked slowly and cautiously through the summer and fall of 1927. On Friday, October 16, three days before the work was scheduled to stop for the winter, Bohlin discovered a specimen that justified the months-long effort. It was an unusually large, wrinkled tooth that appeared to be human. Bohlin personally delivered it to Davidson Black.

Black was ecstatic. The find provided, he believed, strong support of his previous statements about early man in east Asia. He announced the discovery of this third tooth to the scientific world and formally named its original owner, *Sinanthropus pekinensis*, Peking man.

That winter Black traveled to Europe and America to publicize his find. He carried a gold watch chain specially fitted to hold the prized tooth. Though somewhat ostentatious, the chain allowed him to keep his precious specimen with him wherever he went.

Despite Black's air of assurance, many of his fellow anthropologists considered that his classification was based upon inadequate evidence. Dubois' Java man had been more cautiously hypothesized on the basis of much weightier evidence than three teeth. But Black was confident that when diggings continued in the spring, more fossils would be found.

When the excavations resumed in 1928, Bohlin made the first of several finds that provided the evidence Black needed to substantiate his claims. Fragments of a badly damaged human skullcap were picked out of the hill. Black carefully cleaned the pieces and reconstructed a partial skullcap that, in his view, showed a marked similarity to Java man and proved that *Homo erectus* had roamed widely throughout Asia.

Then, on a cold winter afternoon in 1929, Black found his trump card. On December 2, long after the digging had in previous years been suspended, one of the researchers uncovered a skull embedded in a portion of solid limestone. The entire section of rock was brought to Black's laboratory, where the anatomist proceeded to free the skull from its half-million-year grave.

Black worked on the skull for four months. In order to avoid interruption, he labored alone in the dead of night. His routine never varied: he awoke at noon, visited friends and ate at the Peking Hotel, then went to the laboratory in the PUMC anatomy building at midnight and worked until morning. Black's assertions were not universally accepted. The evidence was thought to be slight and, because his theory was dependent on the then dubious claims for Java man, many scientists, Dubois included, insisted that Black was drawing unwarranted conclusions. More than skull fragments were needed if his theories were to be embraced as fact.

In April 1930, the fruit of Black's efforts was displayed: a remarkably well-preserved juvenile skull of *Sinanthropus pekinensis*. Unlike Dubois' *Pithecanthropus* skullcap, Black's specimen was a full skull with the entire region around the ears intact. Black was thereby able to develop a clearer portrait of Peking man than had been possible with the fragmentary remains of Java man. Although there were similarities between the two specimens, Peking man possessed a much steeper forehead, a higher braincase, and greater cranial capacity. These features suggested that

Peking man was younger than Java man—Black put his age at 500,000 years. A convincing description of *Sinanthropus pekinensis* was deduced by analysis of the skull, and Black, Andersson, and their colleagues at the Cenozoic Research Laboratory were left with no doubt that the skull belonged to a true man, not a giant gibbon, as Dubois had called his man from Java.

At a height of about five feet, Peking man was smaller than modern man. But he was not dwarfish and by no means supported the theory commonly held in the early 1900s that the first humans were pygmies. With an undersized skull and a low, flat forehead that retreated from a pronounced, bony ridge along the brow, Peking man looked less like an ape than a modern human with a poorly filled-out head. His jaw was fully developed and protruding, but his chin was small and weak. He had more and larger teeth than modern man. The lower teeth were flat, the uppers, particularly the canines, were projecting and had cutting edges.

From the materials found with the *Sinanthropus* remains, it became apparent that Peking man camped under projecting rocks and in small caves. (Five hundred thousand years ago the region was a dense forest populated by roving bands of animals, and he was forced to confront his natural enemies with weapons fashioned from quartz.) Some crude chopper tools were uncovered but it was difficult for excavators to distinguish the quartz artifacts from the surrounding rock.

Peking man put his primitive tools to good use against the hordes of animals, but his supreme weapon was fire. Layers of red ash found with the fossils gave evidence that Peking man used fire for protection and heat, and that he knew how to manage it and keep camp fires burning for some time. He needed fire not only to hunt and cook deer, wild boar, and to a lesser extent horse, buffalo, and rhinoceros, but also to protect himself from saber-toothed tigers, elephants, hyenas, bats, rodents, and countless other predators that vied with him for the land. The bones of carnivores, especially hyenas, found in abundance among the Choukoutien remains indicated that Peking man was either openly challenged by those animals or that they took over the area when he left.

One chilling deduction was that Peking man was a cannibal and a head hunter. Diggers at Choukoutien found badly damaged bones and bone fragments, which suggested that Peking man was his own worst enemy and that he refused to join his fellows in a community against the elements. Cracked femurs were found, providing evidence to the anthropologist that the bones were split open so the marrow could be extracted, a common sign of cannibalism. The overwhelming number of skull fragments hinted that the skulls had been deliberately cracked open. It is the custom of head hunters to open the skulls of their victims and eat the brain matter, which was believed to contain miraculous powers.

Despite such grotesque behavior, Peking man represented a major step in the evolution of man and his mastery over the elements. He fashioned rudimentary tools and weapons, was proficient in the use of fire, and perhaps even began to develop primitive patterns of speech. He was no longer a tropical vegetarian like the apes, but a rapacious, predatory creature who could adapt himself to his surroundings or travel in search of a different environment.

The wonder of Peking man as an anthropological specimen increased with each new digging season. More fragments were found each year, and each new specimen was used to corroborate evidence from previous finds. Davidson Black continued his intense schedule even though his health began to fail. Friends advised him to leave North China and its severe climate, but the Canadian would not hear of it, and each night, in the deep stillness of the laboratory, he pored over newly discovered fragments. He continued for four years, carefully reconstructing skulls and preparing specimens using a dental drill he had ingeniously adapted to blow a steady stream of dust that gently removed thousands of years of encrusted dirt from the bones.

By March 1934, Black's colleagues and friends were well aware that his health had severely deteriorated. He had inhaled large quantities of the dust he aimed at the fossil fragments, and was experiencing severe respiratory problems. On the morning of March 15, his secretary came to work and found Black slumped over his notes, dead of a heart attack at age fifty. In his hand he held a skull of his beloved Peking man.

Davidson Black had brought worldwide attention to the Choukoutien digs and had elevated the Cenozoic Research Laboratory and the Peking Union Medical College in the anthropological community. He would be missed but the work had to continue. His research and, if possible, a measure of his drive, would be sustained by capable hands.

The torch fell to Professor Franz Weidenreich, a well-known anthropologist who had done significant work on fossil finds in his native Germany. Upon his arrival in Choukoutien, the white-haired professor assumed full supervision of the research and began a detailed study of the fossils.

By 1935, Peking man was the most impressive and by far the most extensive collection of *Homo erectus* specimens ever unearthed. The excavations had produced 14 skulls or skull fragments, 14 lower jaws, a number of post-cranial bones, and 147 teeth. The bones were at first thought to represent 32 individuals; that figure was later raised to 40. Specimens from the trunk portion of the skeleton had not yet been uncovered but the material on hand was overwhelming, and the fossil-rich caves were by no means fully worked.

In 1937, the political situation in China forced Weidenreich and his colleagues to suspend the Choukoutien digs. The Japanese occupied large portions of the mainland and were inching their way toward Peking. The United States was not yet at war with Japan, so the laboratories at the Peking Union Medical College, an American-supported institution, were still safe. But nobody knew how long that condition would last.

Weidenreich and his staff began to discuss the necessity of removing Peking man from the Japanese threat. They first considered hiding the fossils in a vault in either the college itself or somewhere in Peking. A second suggestion was that the fossils be secretly shipped to a peaceful part of China, such as Canton where the Geological Survey of China maintained a station. The third alternative, considered by far the safest, was to ship the fossils out of the country. There was, however, an agreement between the Chinese government and the Rockefeller Foundation that nothing found at Choukoutien was to be removed from the

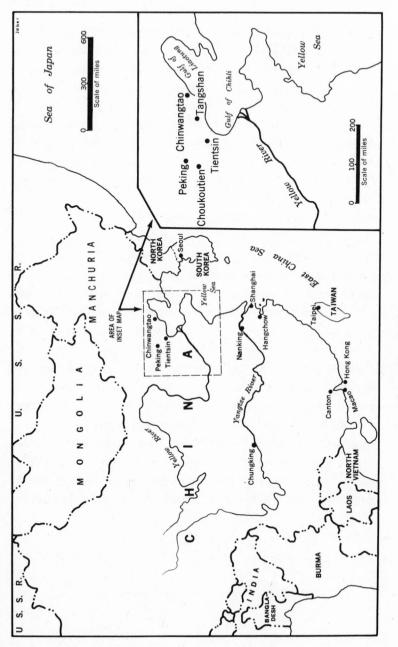

The coast of China.

country. Weidenreich attempted to persuade officials of the Rockefeller Foundation and the United States Embassy in Peking that the fossils should be shipped out of the country as official baggage, but they were unwilling to violate the pact with the Chinese.

The discussions dragged on into 1941, while the fossils remained in the Cenozoic Laboratory. Weidenreich decided to leave the country while he still could. It was suggested that he take the fossils with him to the United States but that suggestion was rejected.

In July 1941, Weidenreich wrote to a colleague: "We arrived at the conclusion that it involved too great a risk to take the originals as part of my baggage. If they were discovered by the customs control in an embarkation or transit point, they could be confiscated. In addition, it had to be taken into account that the objects are too valuable to expose them to an unprotected voyage in so dangerous a time." When Weidenreich embarked for the United States in July 1941, he carried with him casts, drawings, photographs, and voluminous data from the seven years he headed the work at Choukoutien. Two years later he published *The Skull of Sinanthropus Pekinensis*, an exhaustive, definitive study detailing the most minute characteristics of the fossil skulls. Franz Weidenreich had safely reached America, but the bones he and Davidson Black unearthed from a half-million-year grave were left behind in China.

CHAPTER 2
Claire Taschdjian

She was not well known at the Medical College; she had come to PUMC in October 1940 and worked as a laboratory secretary for slightly more than a year. When the order to pack the fossils was issued, however, she was one of the few employees left. She and a Chinese laboratory assistant were assigned to the task of packing and crating Peking man in late November 1941, and that is how it happened that Claire Taschdjian became one of the last known individuals to touch the fossils.

She had come to China from her native Germany in 1934. The shy twenty-year-old had accompanied her father, a doctor, to Nanking where she pursued her love of biology, in the hope of becoming a doctor herself. She lived and studied in Nanking for six years before traveling north to Peking to take the secretarial job at the Cenozoic Research Laboratory. It was not a routine, paper-shuffling position; Claire Taschdjian got closer to the Peking man fossils than anyone except the research scientists themselves.

She worked directly under Franz Weidenreich, who quickly became her teacher and inspiration. What impressed her most about the seventy-year-old anthropologist was the intense yet disciplined manner with which he approached his work. From her first day on the job, Claire sensed that Weidenreich was pursuing the most important research of his life, that the collection of fossils that lay before him was a landmark in the study of physical anthropology.

When Claire came to work at PUMC, excavation had been suspended because of the Japanese occupation. She had missed the excitement of the various discoveries, but she was to become privy to the definitive laboratory research conducted by Weidenreich, Pierre Teilhard de Chardin, and others. The work of classifying and documenting the discovery flourished though the digging had ceased.

Claire soon became indispensable to the researchers. Fluent in English and Chinese, she served as a translator for the research team. But she did more than paperwork. With the Chinese research assistant who later helped her pack the fossils, she made numerous solid and hollow casts of the skulls, the postskeletal fragments, and the separate chips and pieces. Then they labeled, described, photographed, and categorized the casts of the 175 fossil fragments that had been collected.

In the fourteen months that she worked with Peking man and the mountains of research data related to him, Claire developed an emotional attachment to the project. Franz Weidenreich impressed her more each day, and to her respect was added affection. She also admired the expertise and dedication of the Chinese researchers, and realized that this was essentially a Chinese project, one that was immensely important to China and its heritage.

Claire came to know the fossils incredibly well; her knowledge echoed the devoted scholarship of her superiors and the pervasive belief that the project was characterized by scientific research of the highest quality. The extensive collection was not the result of a chance discovery or a lucky break. A visit to the excavation site—the mountains of earth that had been sifted so meticulously by hand—would testify to the magnitude of the digs.

When Weidenreich left China in 1941, Claire was sorry to see the end of her seven-month relationship with her friend and teacher. He had taken her into his confidence, and he had shown her first-hand the techniques of physical anthropology. He had not finished writing his massive study on the skulls, the last portion of the collection to be described, but the Japanese menace was too great to ignore any longer, and Weidenreich would have to complete his work at the American Museum of Natural History in New York. His departure also ended fourteen years of research in the Cenozoic Laboratory. Teilhard de Chardin and the handful of Chinese researchers remained, but in the latter part of 1941 even these men would be unable to work on the fossils. That would be left to Claire Taschdjian.

She hated the Japanese invaders, as did the others in the laboratory, but throughout her first years she tried to work on as usual. The atmosphere was tense; the researchers and PUMC officials lived in a state of uncertainty concerning further Japanese incursions. It was naive to hope that science was immune to the chaos of war.

In late November, the Chinese Nationalist government in Chunking approved plans to remove the fossils from the country. PUMC officials asked Claire to prepare the specimens for the trip because she was the only member of the research team who remained. For three days Claire worked in the vacant Cenozoic Laboratory. The fossils covered every available surface in the lab. Each was checked against an inventory list, then wrapped and placed in small cardboard boxes. These in turn were packed in one of two redwood crates that measured about twenty inches by twenty-four inches and were labeled "A" and "B." In addition, the sides of each crate were stencilled with the initials PUMC. One was to be packed with *Sinanthropus* remains, the other would contain specimens from the site designated as the Upper Cave. Claire Taschdjian's inventory list for the Peking man box included 5 skulls, about 150 jaw fragments and teeth, 9 thigh bones and fragments, 2 upper arm bones, a collar bone, and a wrist bone.

The contents of the second box interested Claire as much as Peking man. The Upper Cave remains were fossils of a much more recent individual, perhaps 10,000 to 20,000 years old, who greatly resembled the structure of modern man. The Upper Cave bones definitely evidenced a primitive Oriental race, forerunners of the modern Chinese. But because it was found in the shadow of Peking man, little research had been done on the material.

Weidenreich had at one time told Claire that the Upper Cave fossils were scientifically more important than Peking man as long as they remained unexamined. He told her that since Peking man had been thoroughly discussed and set into casts which showed every groove, its scientific value had been established. But the Upper Cave material, he told her, was the earliest evidence of *modern* man found outside of Europe. Little was known about these remains.

The booty from the Upper Cave was small: 5 complete skulls, and vertebrae belonging to at least 8 men and women. They were packed with all the care given to Peking man and locked away in box "B."

Both boxes were padlocked and transferred without

ceremony to the vault in the college strongroom. That was the last Claire saw of them.

Claire remained at PUMC as secretary to the comptroller once the task in the laboratory had been finished. A week after she had packed the boxes, a United States Marine officer came to the college and gave her the two padlock keys. He casually remarked that the mission had been accomplished. Claire assumed he meant that the boxes had been safely delivered to the marines and everything was in order.

A few days later, the long-anticipated enemy assault took place. PUMC was thrown into complete disarray. The Japanese took over the campus and detained the director, the comptroller, and other college officials as civilian prisoners of war. Because she was a German citizen, Claire was considered a "friendly alien" and allowed to pursue her studies in Peking.

Before she left the PUMC campus, she watched the Japanese take over Lockhart Hall, the site of a large vertebrate laboratory, and discard specimens as if they were rubbish. Chinese peasants were quick to collect the remains and carry them off. Claire later learned that many of the Lockhart Hall specimens turned up in thieves' markets in Peking, and she recalled that it is said in China: Nothing is ever thrown away, nothing is ever lost.

Throughout the war Claire stayed in Peking, coming and going as she pleased. In 1942 and early 1943, the Japanese secret police, the Kempeitai, questioned her frequently about Peking man, but she could tell them no more than what she knew about the specimens while they remained at the PUMC laboratory. She could only say that she had touched the fossils and that she had known them well; she acknowledged that she could easily identify them if they were found. But she despised the Japanese and could not hide her resentment of the Kempeitai's questions. If it had not been for the Japanese, her beloved Franz Weidenreich and the precious fossils would have been safe in the Cenozoic Research Laboratory, and the work would be continuing.

CHAPTER **3**

Herman Davis

He had arrived on the last troop ship to China on September 25, 1941, and was sent to the marine camp in Tientsin. Pharmacist's Mate Third Class Herman Davis was short, stocky, and musclebound. He spent his leisure time weight lifting and wrestling, and often posed for photographs wearing swim trunks, raising his arms in "Muscle Beach" stances which silhouetted the carefully developed ripples along his arms and thighs. But in Tientsin there was little time for posing and few women to impress, and Davis quickly settled into his job as pharmacist's mate under Lt. (jg.) William E. Foley, M.D. He respected the thin, serious-eyed doctor and looked to him for advice and instruction. One day Dr. Foley would realize that the best thing that ever happened to him was his acquaintance with Herman Davis.

Davis stayed at Tientsin until Thanksgiving when word came at Camp Holcomb in Chinwangtao that one of the marines had come down with gonorrhea. The stricken soldier was sent south to Tientsin for treatment and Davis went north to replace him. He took charge of the small medical department at Camp Holcomb, still under the direction of Dr. Foley.

While Herman Davis was making his way to Chinwangtao, the PUMC redwood crates were transferred from the
college vault to a waiting car which drove them the short
distance to United States Marine Headquarters in Peking.
The crates arrived in the office of the headquarters command at a time when Col. William W. Ashurst had other
things on his mind. The marine commander was surrounded
by a hostile army and had no time to worry about a
shipment of bones, no matter how valuable they might be.
He ordered the collection transferred from the redwood
crates to regulation marine footlockers and consigned them
to the care of William Foley, who was scheduled to return
to the United States as soon as the North China marines
evacuated China.

When Foley was informed about his additional baggage
he made a mental note to give it special attention. The
footlockers were to be added to the cargo of a marine
transport train bound for the port city of Chinwangtao,
where the gear would be stored in the marine outpost,
Camp Holcomb, until a troop ship arrived.

A handful of marine embassy guards were ordered to load
and accompany the military cargo on the transport train. It
was a tedious detail, but it was their last before leaving the
cold of China for the warmer Philippines. As they loaded
the footlockers, they were alerted to those bearing the
names of Colonel Ashurst and Dr. Foley. They assumed that
the lockers contained personal property; it was common in
the service to give special attention to whatever belonged to
the brass.

The guards were also told that some of the footlockers
were special cargo from the medical college, a collection of
bones called Peking man. Though they knew little about

fossils or anthropology or how important the contents of the footlockers actually were, the guards carefully loaded the designated lockers and dutifully kept an eye on them.

The old steam train jerked and swayed across the countryside for three days before it reached Camp Holcomb, a mere 140 miles from Peking.

While the train and its marine cargo traveled toward Chinwangtao, the U.S.S. *President Harrison* steamed across the Yellow Sea. The *Harrison* was a private passenger liner converted for use by the navy; its first mission was to evacuate the North China marines from the mainland. By late November, the *Harrison* was still hundreds of nautical miles away. Barring any difficulties with the Japanese, it was scheduled to land in Chinwangtao on December 11, 1941.

The marine train pulled up to Camp Holcomb and unhitched three boxcars full of gear onto a railroad siding which ran into the camp. One of the handful of embassy guards was ordered to stand watch. As he did, he was approached by a Japanese army officer and another Japanese dressed in civilian clothes. They asked the guard what was in the boxcars and the marine answered that he had no orders to disclose such information. The Japanese persisted and asked if they could check the contents of the cars themselves. Again, they were refused and they finally relented, leaving the marine to wonder about their curiosity.

Shortly after his arrival, Davis had received a radio message from Dr. Foley asking him to take charge of the footlockers on the train. Davis met the train and, with the help of a few other marines, he unloaded more than two dozen lockers with "W.T. Foley, USMC" stencilled on the sides, and carted them to his room. Davis did not see any lockers belonging to Col. William W. Ashurst.

Herman Davis and the seventeen other marines at Camp Holcomb were housed in a single brick building, a one-story structure sectioned into four large sleeping areas. Across from the barracks was a mess hall with a kitchen and a small recreation area. To the south a few solitary shacks had been built for visiting officers. The most conspicuous sight in the camp was the railroad siding on which stood boxcars filled with ammunition and supplies. The boxcars were not well guarded, despite the presence of the Japanese. The camp had only one man on watch, and he patrolled the whole camp.

Davis had been given the entire north end of the barracks for his own quarters, a four-bed infirmary, and a laboratory. There was plenty of room for Foley's footlockers. He piled them all over—under beds, against a wall, in the corners. They were in easy reach and made convenient poker tables for some of the men just down from Peking or Tientsin who wanted to relax and play a few hands. Davis was never told what was in the various footlockers and didn't ask.

The footlockers remained in the medical corpsman's room through the first week of December. He and the seventeen other marines paid little attention to them; they were more concerned about what was going on in the rest of the world. The Japanese threat to the United States appeared imminent, and the tension was heightened on December 4 when the North China marines received orders from Washington to destroy their secret codes and all coding material. The marines counted the days until they could get out of China and head for the Philippines.

Davis was relieved when Sunday morning came and went. Military men generally believed that if the Japanese were going to invade they would do it on a Sunday morning, when American soldiers customarily dried out from the carousings of the night before. Davis enjoyed the rest of the day, confident that he would soon be in Manila. That night he played a few hands of poker, again using the doctor's footlockers, and turned in. It was a clear, bitter-cold evening, with temperatures plunging to forty degrees below zero. Davis piled his bunk with as many blankets as he could find and slept deeply.

Things had changed dramatically by the time he awoke Monday morning. Davis was roused by shouting and confusion in a camp that was swarming with Japanese soldiers. There were hundreds of them, some in ditches, some squatting behind marine cars, others almost ludicrously trying to hide behind the four-inch-diameter trees that dotted the camp. Davis heard the drone of fighter planes overhead, and in the bay about a thousand yards to the south of the camp, he spotted a Japanese cruiser.

It was still early morning and despite the bedlam and the threatening presence of the Japanese, Davis decided to start off his morning as he always did. He dressed and shaved and made his way over to the mess hall. In the face of the storm he knew was to come, Herman Davis sat down and enjoyed a leisurely breakfast of pork chops.

When he returned to the barracks, he learned what had happened. Though Camp Holcomb had made it through Sunday without an invasion, a base on the other side of the world, where it was still Sunday, had not. The Americans were formally at war with the Japanese, in Chinwangtao, China, as in the rest of the world, on December 7, 1941. On orders from their platoon sergeant, the senior enlisted man in the camp, the marines began to mobilize weapons and ammunition for a possible defense of the camp. Though there were only eighteen of them, the marines were prepared to take the Japanese challenge head-on. The thought of surrender in the face of overwhelming odds never entered their minds. They were United States Marines and now was the time to prove what that really meant. They hustled about the camp readying their personal arms, rifles, and pistols, and unpacking the heavier machine guns and antitank guns.

A man was sent to retrieve Lt. Richard Huizenga and CWO William Lee, who had left at 6:00 A.M. that day to go duck hunting. The two had come up to Camp Holcomb from Tientsin for the hunting season, staying in the small visitors' shacks behind the mess hall. Huizenga and Lee returned through Japanese lines as Davis and the others were unpacking the heavy weapons. Lieutenant Huizenga immediately took command and joined the furious preparations. It was the moment he had been trained for; he was ready for the fight.

Davis went to his room and began checking his first-aid equipment and supplies. He knew he would need all he had; the marines stood little chance against the horde outside. He decided the Red Cross insignia on his arm would be of small help to him so pulled it off, and joined the others in the shower rooms where they were hurriedly washing cosmoline, the thick, greasy preservative, from the heavy guns. He returned to his end of the building and pulled everything he owned, including Dr. Foley's footlockers, in front of the doors and windows. Then he propped up his machine gun and waited.

It was an eerie time for Davis, sitting nervously behind a weapon he had never fired before, looking at the hundreds of Japanese faces outside the window. He had the gun

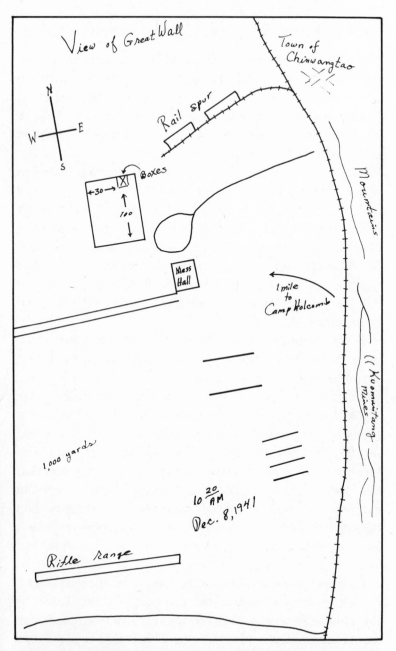

A hand-drawn map of Camp Holcomb. This camp was in view of the Great Wall of China and was the last outpost of the marines in North China up to December 8, 1941.

primed; he figured he would last about ten seconds. He did not fear dying, perhaps because he had been trained to expect it and had gained a sense of the inevitability of death that a civilian never possesses. His only thought of death was that it was a shame that he had not left a child back in the United States to carry on his name.

At the other end of the barracks, Lieutenant Huizenga met what he considered the first challenge when he and a few others captured four soldiers sent into camp under a white flag to demand the Americans' surrender. The marines sat with their four prisoners and waited for word from Colonel Ashurst in Peking. The order to surrender came at 10:40 A.M., and the short-lived but determined defense of Camp Holcomb ended peacefully; not a shot had been fired.

Minutes later, Davis was back in his room deciding which of his possessions to take with him in the single footlocker permitted by the Japanese. He thought immediately of the weather and pulled out a bag of his warmest clothing. He grabbed some cigarettes, some medicine, and had room for little else. The Japanese assured the men that they would receive the rest of their belongings later on, but few believed that promise would be kept. They had heard that the average Japanese soldier was obsessed with American gadgets and had a special weakness for watches and cigarette lighters. They counted on the Japanese to take a lot of souvenirs.

Davis' capture was a sudden, largely unforeseen event. For Dr. Foley in Tientsin, the prospect of confinement by the Japanese was something he had contemplated and feared for months before the attack on Pearl Harbor. Foley had come to China in 1938, a year after he graduated from Cornell Medical School. Although he was assigned to the Marine Medical Corps, he was also able to teach in medical colleges in Hong Kong and Peking.

Foley's duties allowed him to travel widely throughout China. He became fluent in several Chinese dialects and made a number of close Chinese and European friends in various towns and villages. Most of his first two years on the mainland were spent in Hong Kong. He was transferred to Peking in May 1941 as head staff medical officer for the North China marines. His personal quarters were in the compound in Tientsin, but he made frequent trips to Peking, both to the marine headquarters and to the Peking Union Medical College, where he held classes.

While teaching at PUMC Foley visited the Cenozoic Research Laboratory and with the permission of laboratory personnel viewed the awesome collection of fossils that had made headlines ever since their discovery fifteen years earlier. The fossils did not become a personal concern to Foley until Colonel Ashurst asked him to make sure the specimens arrived safely in the United States.

When Monday, December 8, arrived, Foley was by no means resigned to the inevitability of becoming a Japanese prisoner. The thin, dark-haired doctor tried to convince marine officers in Tientsin that they could not be taken prisoners of war because of the Boxer Protocol, a diplomatic agreement that stated that in the event of war between signatory nations (Japan was one of these), all members of foreign armies would be permitted to return to their home countries with diplomatic status.

Foley persuaded a superior to confront the Japanese with the terms of the protocol. The Japanese commander, in turn, forwarded the question of diplomatic status to his superiors in Tokyo. For a few hours, the marines held hopes that they might be permitted to leave China after all. But by noon, when a general order had not come through, the Japanese took the Tientsin division prisoner. The marines were ordered to remain in their compound where they would be joined by fellow soldiers from Peking and Chinwangtao.

Foley, however, got a temporary reprieve. The commander of the North China Japanese regiments knew Foley from various diplomatic receptions and put through an order that the doctor be allowed to continue his work at the hospitals in the area. Foley was also permitted free passage through Japanese lines.

Later that day, the trainload of marines from Chinwangtao arrived in Tientsin and Foley was reunited with Herman Davis. Foley didn't ask about the footlockers; Davis, like all the others, had arrived with one bag of personal belongings and nothing else. A week or two later another trainload of marine baggage came in from Camp Holcomb. The gear was dumped into a pile in the Tientsin camp and each marine picked through it piece by piece to find his belongings. Some of them got most of their gear back; others retrieved only a few things. Nobody was really sure if the pile contained everything that had been in Camp Holcomb or if the Japanese had stolen or discarded some of the goods.

Foley received most of his footlockers intact. He reasoned that his belongings had not been thrown in with the enlisted men's baggage because of the Japanese regard for rank. But when Foley opened his boxes, he discovered that a number of skulls he had used as teaching aids were missing. He also noticed that many valuable mementos given to him by Chinese friends were gone. Foley apparently did not open the footlockers assigned him by Colonel Ashurst, those containing Peking man.

Since he was not sure how long he would be granted special status by the Japanese, he decided to store his footlockers in Tientsin. He left some in a warehouse, operated by a Swiss firm in Tientsin. He sent some other material to the Pasteur Institute in the French concession in Tientsin, where he had once worked. Other goods were forwarded to various personal friends upon whom he could rely. But he told no one, not even Herman Davis, the specifics of his arrangements for the storage.

A few days later, Foley and the other marine officers lost their privileged status and were interned with the rest of the marines in the Tientsin compound. They passed the month of December, celebrating a bleak, uninspired Christmas. Colonel Ashurst and his Peking regiment were transferred there on January 10, 1942, bringing the entire marine contingent together. They were then shipped out to the Woo Sung prison in Shanghai, traveling the long distance in boxcars, talking among themselves, trying to keep spirits from sagging.

Once in Shanghai, the marines settled into the task of keeping up morale and preventing the Japanese from confiscating what personal belongings they had managed to take with them. The enlisted men like Herman Davis held on to cigarettes and watches. The officers tried to conceal personal papers and valuables they had brought with them.

The Japanese searched the American prisoners daily. They would be marched into an open field and ordered to spread out all their belongings for inspection. A favorite ploy of the prisoners was to dig a hole and bury what they wanted to conceal. Then they would spread the rest of their gear over that spot and hope the Japanese would not look too closely. Once the search was over, they dug everything up. Each day this ritual was repeated, and the Americans were able to keep personal mementos hidden from the Japanese for months.

As the war progressed into its second and third years, the American prisoners devised ways to withstand the psychological pressures inflicted by the Japanese. The Japanese propaganda machine attempted to portray American and British prisoners as happy and satisfied, always willing to denounce Allied forces and the Allied cause. In fact, some prisoners attempted to confront the Japanese openly with a barrage of rebellious tactics; others attempted to escape.

Foley played a part in the escape attempts by writing notes in Chinese to aid the escapees in their flight. His role was discovered when escaped prisoners were caught with his notes in their possession. To punish the doctor, the Japanese locked him in a small wooden shack for as long as thirty hours at a time. The solitude was maddening, but Foley survived his imprisonment, partly due to the efforts of Herman Davis.

The isolation shack stood at the end of the prison's barracks area. It was unheated, nearly totally dark, and not high enough to allow a man of average height to stand. A Japanese guard stood watch around the clock. To get to it, one had to pass the washing racks outside the barracks and then go through a lavatory building. The Japanese used large, earthenware jugs sunk in the ground for latrines. Small booths were built over the jugs and a hole cut in the floor of each booth, over which the prisoners squatted. The floor stood a few feet above the ground; small doors were built in the side walls to permit access by prisoners assigned to clean out the jugs.

Whenever he could, Davis crept in by one of the doors, passed over the fetid jugs, and managed to crawl undetected to the rear of the shack where Foley was imprisoned. Through a loose slat, he slipped Foley bits of food, drink, medicine, and almost anything else the doctor needed. When Foley was released, he appeared almost as strong and healthy as when he went in.

In 1944, the marines were moved to the Kiangwan prison camp on the outskirts of Shanghai. The camp was located between two airports, and the Americans soon realized that by noting the coming and going of Japanese planes, they could build a valuable store of intelligence information for the Allied forces. They were able to get the information out of the camp with the help of Chinese nationals, disguising the operation to look as if they were exchanging money and goods with the Chinese through the prison fence. The Japanese became suspicious when American planes began to attack Japanese aircraft at strategic times.

Foley was immediately suspected because of his knowledge of Chinese. As the strikes continued to cripple the Japanese, they suspected Foley more and more. When Foley refused to admit his guilt, the Japanese resorted to torture. They began with the Spanish windlass: a string is tied around a finger and the back of an arm, then slowly tightened until the finger breaks. Foley kept silent.

The next ordeal was the water torture. Foley was tied

down on his back while large pitchers of water were poured over his face. Over an excruciatingly long period of time, he choked on the water, inhaling much of it.

Davis was close to Foley through all of this. As a medical corpsman, he was assigned to handle those who died in camp. He performed postmortems and sealed the bodies in boxes for burial. During Foley's torture, Davis was stationed outside the torture chamber routinely awaiting a new corpse. In this case, however, he had a special concern for the victim.

When the Japanese officers finally walked out of the building, they were apparently convinced they had killed him. Davis hurried inside and saw Foley lying on the floor. He was sure the doctor was dead until he felt his wrist and detected a slight pulse. Foley was in a deep coma and had little color. Davis immediately gave the doctor an injection of caffeine and the stimulant sodium benzoate. He then called out to another pharmacist's mate, who helped him carry Foley on a stretcher back to the infirmary.

The two men worked on Foley through the day and into the night. He was catheterized and fed intravenously from time to time but he remained in a deep coma. They worked without interruption, refusing to lose hope, even though Foley showed no signs of recovery.

On the fifth day of treatment, Davis began another catheterization. Suddenly, he was interrupted by his patient.

"Let me do this," Foley said, looking up at Corpsman Davis.

Foley's further recovery was rapid. Though he had returned from the dead, the Japanese did not torture or punish him again. Like other prisoners, however, Foley was approached by an interpreter to sign a confession of guilt to be used for propaganda purposes. Again he refused.

Months later, Foley and Davis were shipped out of Kiang-wan to a prison camp in Fungtai, near Peking. From there they were sent to Manchuria, Korea, and ultimately to an iron mine in northern Japan. During this time, Allied forces had reversed the tide of the war in the Pacific and delivered decisive blows against the Japanese. Davis and Foley had their own personal victories. They had spent the war in China together, prisoners of the Japanese. In 1945, the two of them came out alive and well—and inseparable.

CHAPTER **4**

The Search Begins

The possibility that the fossils were lost or destroyed once they were put in the hands of the marines was a real one. In the aftermath of Pearl Harbor and the end of American military presence in China, the Japanese occupational forces plundered the northern provinces. Several months after Foley and Davis were sent to Shanghai, a contingent of Japanese arrived at the PUMC laboratory and confiscated whatever relics they could find.

The Japanese combed the Cenozoic Laboratory and recovered some specimens—mostly fossil fragments, remnants of stone tools, and nonhuman remains—from the Upper Cave collection. They also looked in the college's safe but reportedly found nothing. The Upper Cave artifacts were taken to Tokyo University. Dr. Hisashi Suzuki, then head of the school, later explained that the remains were taken "on loan" from the Peking laboratory.

The specimens were returned to Peking in 1945 by Frank Whitmore, a United States Army geologist, but not before Whitmore had excited the scientific world by announcing that he had located Peking man. Whitmore wrote to colleagues in America that he had found the original

Sinanthropus fossils in Tokyo University and was soon to return them to the Cenozoic Laboratory. But when Whitmore met Dr. Suzuki, the Japanese professor denied any knowledge of Peking man and instead produced the Upper Cave materials. It can be assumed that those fragments had been left behind in the laboratory because PUMC officials did not consider them valuable enough to pack along with Peking man.

Although Whitmore returned the Upper Cave remnants to the Cenozoic Laboratory, he was for years afterward suspected of knowing more than he had revealed about the whereabouts of Peking man. He vehemently denied this and said he believed the collection had been lost at the bottom of the Yangtze River.

When the Upper Cave remains arrived in Tokyo in early 1942, the minister of culture was informed that Peking man was missing. This report was passed on to Emperor Hirohito, who ordered the North China Expeditionary Force Headquarters to take necessary action to recover the fossils.

Japanese agents zealously pursued their quarry. They thoroughly examined the United States Embassy in Peking, but without results. The Kempeitai fiercely questioned PUMC officials who were, at the time, civilian prisoners of war.

The most severely harassed was Trevor Bowen, PUMC comptroller. In April 1943, he was incarcerated for five days in a wooden cage too small to lie down in and fed only rice and water. He gave the Japanese no information and finally convinced them he knew nothing about the fossils' whereabouts. In July or August, the Kempeitai took Claire

Taschdjian to search through United States Marine Corps belongings stored in the Bryner and Co. warehouses. Again the Japanese found nothing.

"I suggested to these people early in the game that the military in charge of the looting at Chinwangtao should be questioned," Claire said. The Kempeitai, however, continued with their interrogation of the PUMC staff.

The intensity of the Kempeitai investigation led the PUMC prisoners to surmise that if the Japanese troops who had taken over the Chinwangtao camp had indeed destroyed or lost the fossils, they would have been too scared to report it to their superiors. Regular soldiers may have feared treatment even more harsh than the harassment suffered by the civilian captives.

At the end of the war, American military personnel also made a search for the fossils, largely at the request of Trevor Bowen. Bowen told marine investigator Albert Scalcione that he had delivered three *boxes* to Colonel Ashurst at the marine barracks in late November 1941, and he supplied detailed diagrams of the boxes. Perhaps because of his own confinement and torture, Bowen was a zealous searcher, a dramatic change from the early days at PUMC when, according to those who were there, Bowen took no special interest in the fossils.

With the help of Bowen and a Japanese investigator who spoke fluent English, Scalcione questioned almost all of the PUMC staff. He was unable to learn any more than the Kempeitai had in 1943.

While Scalcione made the rounds in Peking, two other marine lieutenants conducted searches in Tientsin and Chinwangtao. Toward the end of the war, American planes bombed Chinwangtao and several warehouses were burned to the ground. The possibility that Peking man was reduced to ashes could not be denied.

Colonel Ashurst, who died a few years after the war, filed this report following the United States Army's official search on March 18, 1947.

During November 1941, several boxes were accepted by me from officials of the Peking Union Medical College for shipment to the United States. These boxes were shipped together with other property belonging to the Marine Detachment, Peking, China, via rail to Chinwangtao, China, in late November or early December in freight cars guarded by U.S. Marine Corps personnel. These materials remained in the cars at Chinwangtao awaiting shipment in the U.S.S. *President Harrison* to Manila, Philippines, and were so located when the war started on December 8, 1941.

I have been told by an officer recently on duty in China that a search was made in Chinwangtao for the Peking Union Medical College materials since the end of the war. To the best of my knowledge, no supplies which were at Chinwangtao at the beginning of the war have been recovered.

Ashurst's statement and the negative findings of the Japanese and the Americans—particularly Scalcione—pointed in the direction of the marines in charge of the baggage and to the U.S.S. *President Harrison* which was to have taken the fossils aboard.

Though hardly a mystery ship laden with treasure, the *Harrison* was never completely absolved of its connection with Peking man. In 1972, officials of the American President Lines, the company that owned the *Harrison*, conducted an inquiry into its fateful trip to China thirty years before. Some of the company's employees remembered the operations of the *Harrison* in 1941 and were eager to help determine whether or not it could have taken the North China marine cargo aboard.

According to Capt. John P. Chiles, Director of Offshore Operations for the President Lines, the *Harrison* sailed from San Francisco on October 17, 1941. It steamed via Suva, Port Moresby, and the Torrez Straits before arriving in Manila on November 12. Two days later, President Roosevelt ordered American soldiers out of mainland China. The *Harrison*, along with the U.S.S. *President Madison*, was immediately chartered by the navy as a troop transport, and both ships sailed for Shanghai at the end of November. The *Harrison* discharged her commercial cargo at Shanghai on December 4, and was ordered to Chinwangtao.

In a letter to American President Lines dated September 12, 1972, Chiles wrote,

. . . Immediately after the bombing of Pearl Harbor, the *President Harrison* was prevented by the Japanese from embarking the U.S. Peking Marine guard and their baggage at the Yellow Sea (Gulf of Liaotung) port of Chinwangtao.

At the time [of the Pearl Harbor attack] the ship was off the mouth of the Yangtze River approximately 700 miles south of her destination. Her master, Capt. Orel Pierson, succeeded in eluding his captors and ran the ship aground on Shaweishan Island (She Shan), a high rock at the entrance to the North Channel of the Yangtze (Ch'ang Chiang).

Chiles added that those of the ship's officers who held military reserve ratings were confined as prisoners of war at various camps in China and Japan. The Japanese salvaged the *Harrison* and refitted her as a transport. For three years she sailed as the *Kakko Maru* and the *Kachidoki Maru*; she was sunk by a United States submarine in September 1944.

The possibility that the fossils were loaded aboard the *Harrison* is remote. Chiles and the surviving officers who were on board at the time the ship was seized have no knowledge of such a shipment. Furthermore, Col. George Newton, a member of the United States Fourth Marines in 1941, stated, "I was informed of the arrangements which were made to remove the bones from their Peking repository for safekeeping in the U.S. I was the last U.S. Marine courier to travel from Peking to Shanghai in November 1941, and the shipment was not made through that port or on the *President Harrison*."

Still, stories continued to circulate during and after the war. One claimed that the fossils reached Chinwangtao but as they were being transferred to the *Harrison*, the loading barge tilted and the crates slid into the harbor waters and were lost. Another opinion was advanced by a man who claimed acquaintance with Teilhard de Chardin to bolster his credibility, despite the fact that Teilhard's last contact with Peking man was at the Cenozoic Research Laboratory. The theory held that the fossils were loaded onto the *Harrison*, which was scuttled *after* it left port. When the Japanese salvaged the ship, it continued, the fossils were found among the cargo, judged to be ancestral bones worthless to all but the deceased's family, and discarded.

Both stories are unlikely for the simple reason that they depend on the *Harrison*'s presence at Chinwangtao, a port it never reached. All the crew members interviewed agreed that the craft was abandoned 700 miles south of its destination.

Further inquiries about the *President Harrison* led to the same dead end: the rocky island of Shaweishan. Important information about Camp Holcomb and the marine transport to Chinwangtao, however, remained with surviving members of the North China marines. The subject of Peking man frequently comes up at their yearly reunions. Gerald Beeman, curator of the Cleveland Museum of Natural History and formerly a marine at Chinwangtao, has maintained a keen interest in the fossils. He was quick to refute a 1959 article in *Science* magazine which attributed the loss of Peking man to the *Harrison*. In his reply, printed in the magazine in May 1959, Beeman described his experiences in Peking as an embassy guard in 1941, and added relevant details he had gleaned from fellow marines at the various reunions. Beeman pieced together the individual accounts of the journey of the transport train from Peking to Chinwangtao, and added his own theory about what happened to the special cargo.

One of the marine train guards gave Beeman this account of the transfer: "I know the boxes with the bones left Peking and arrived at the rifle range in Chinwangtao. I was in charge of guarding some property we were sending from Peking to Chinwangtao which included the boxes. They were the last thing loaded into the boxcar from a truck before we left the freight yard in Peking." The guard maintained that the transport he was on was the last from Peking to Chinwangtao and the only one which could have included the fossils.

All the same, there were several other transport runs between Tientsin, Peking, and Chinwangtao. Some of these trains were stopped by Japanese troops at the outbreak of the war, and if the guard was mistaken about his train's cargo, the Japanese might have removed the fossils from one of the others.

Assuming the guard's story was accurate, Beeman focused on Camp Holcomb. He believed the most critical question to be asked about the Japanese takeover was whether they had searched the camp or just looted it. He knew that the marines' personal effects were shipped to them at prison camps in Tientsin, but the possessions arrived in a jumbled heap, and it was apparent they had been thoroughly searched. The sorts of things missing from the lockers might provide a clue whether the Japanese were looking for specific items or had merely been careless with marine belongings.

If the camp was looted, Beeman reasoned, the fossils could have ended up anywhere. They might have been thrown into the sea or, if left amidst the debris of the camp, they might have been carried off by Chinese peasants in search of dragon bones.

But Beeman doubted that the Japanese discarded the fossils, since it is clear that they were on the lookout for cultural treasures. Beeman spoke to the marine guard at Camp Holcomb who was approached by the two Japanese who wanted to examine the newly arrived cargo. Their interest convinced him and other marines that the fossils were not discarded by the Japanese. There was, of course, the possibility that they were discovered by foot soldiers not under the supervision of commanding officers. These uninformed men might have tossed them aside or sold them to Chinese traders.

If the fossils were part of American guns and supplies packed in the standing boxcars, Beeman continued, the Japanese might have shipped the entire load away to another theater of operation. The fossils could then have gone anywhere, from Japanese outposts in China to the islands of the South Pacific.

Of the total detachment of North China marines, 18 were

stationed at Chinwangtao and Camp Holcomb, 38 at Tientsin, and 144 at Peking with the United States Embassy and the Legation quarters of Colonel Ashurst. Few if any marines at Tientsin knew anything about the fossils and their transport, since the shipment went from the laboratory in Peking directly to Chinwangtao. A number of others who attended the yearly reunions had stories to tell.

On December 8, Walter J. Reilly was on board a transport train en route to Chinwangtao from Peking. The Japanese stopped and thoroughly searched the train. Reilly wasn't sure the fossils were on his train ("I was in a boxcar that had dozens of boxes"), but he said that if they were, the Japanese could have intercepted Peking man at Tangshan. If they did, Reilly suggested, they certainly knew what they had.

The Japanese lieutenant who took me off the train at Tangshan was a very thorough soldier. The train of five boxcars had been pulled over to a siding in the railroad yards there and he carefully inventoried all of our personal gear, once marvelling that I had four pairs of shoes. He carefully counted all small items and made me count them, six pairs of shoelaces, and so forth. If he had anything to do with the rest of the baggage and the equipment in the rest of the cars, a careful inventory would have been taken.

But Reilly had doubts about the sophistication of the average Japanese soldier; he believed that if the fossils had somehow fallen into the hands of infantrymen, they would have been destroyed, discarded, or sold to the Chinese. His theory also applied to the fate of the fossils if they were confiscated with the rest of the marine baggage. "If the average soldier looting a warehouse where the gear must have lain for at least three years had found an old box filled with cotton batting and a few brown objects, he would have thrown it on the dump pile."

Most marines were only concerned with their own gear and the difficulty they had keeping it once they were taken prisoner. David Timpany, a marine stationed in Chinwangtao, remembered taking his footlocker along to the prison camp at Tientsin and noting that the Japanese didn't seem too interested in American money or valuables. When the marines were transferred to the Woo Sung prison camp near Shanghai, Timpany said, the Japanese opened and inspected the lockers and then marked them with chalk to indicate they had been checked. "But there were several trunks the Japanese never viewed, since the men would switch them from one line to another and chalk the trunks themselves," he said.

Timpany said he was never told what was in the unopened trunks but he'd heard rumors that they contained weapons, or rocks taken from the Great Wall. He also said, "I never knew what was in them, but the fellows who had to carry them for the Peking officers complained that the trunks were very heavy."

Timpany was one of the few marines who said he was able to keep his trunk throughout the war. Some marines said they had stored trunks at the Swiss Consulate in Shanghai before being shipped to prison camps in Japan. Walter Freiberger got his baggage back after the war but was somewhat unhappy—the items he stored were of little value, but the valuable possessions he kept with him had been confiscated. Morris Carson said he also stored a trunk with the Swiss but it had been broken into and much of the contents stolen. The thefts were odd: three silver cigarette cases were left untouched but some wood carvings he had made himself were taken. John W. Whipple wrote that he thought much of the marine baggage found its way to friends of various soldiers in Peking, Tientsin, and Chin-wangtao. He said these people were willing to keep things for the Americans or arrange to have them shipped out.

Kenneth Clark was one marine who didn't believe the boxes containing the fossils could have been moved from one prison camp to another. "When we left Peking for Shanghai, we all traveled aboard cattle cars and then had to walk five or more miles to the prison camp in Woo Sung. There was no transportation for our personal effects. We carried everything we had with us. I don't remember anyone carrying a footlocker or lockers."

But Clark did not rule out the possibility that some marines might have kept their lockers in the Tientsin prison camp. He said there were several footlocker-sized boxes left in the camp when the Americans were transferred to Shanghai.

Col. John A. White believed the fossils did not even make it to Tientsin, but were discarded in the rubble at Camp Holcomb. White also discounted the possibility that a box could have been hidden from the Japanese. "No such box arrived in early August 1945, at Nishi-Ashebetsu, the last officer POW camp, in Hokkaido mountains. I saw every box that the camp authorities opened." But White did concede that some footlockers at Nishi-Ashebetsu had not been opened and they could have been returned to the United States after American forces occupied Hokkaido.

Two other marines stationed at Chinwangtao supplied still another story. Emit Logan and Roy McCarthy cited mysterious events which occurred in the hours just prior to the Japanese takeover of Camp Holcomb. Logan was an orderly assigned to officers' quarters. He had orders to awaken Richard Huizenga and William Lee early on December 8 so the men could go hunting. He did so, and the pair left and returned, according to Logan, shortly before the surrender. But he wondered if their mission hadn't been something other than duck hunting.

As McCarthy told it, "On the day of December 8, two of our officers at Camp Holcomb, Lieutenant Huizenga and Warrant Officer Lee left camp using one of the trucks. There was never any explanation of why they didn't return until shortly before our surrender on December 8."

Logan also recalled that he became an orderly for the late Comdr. Leo C. Thyson. He said Lee visited Thyson's room frequently and always after a personal inspection by the Japanese. "Lee's standing remark was that he had been given a reprieve. When I would ask him what he meant, he would tell me that he had a valuable shotgun in his locker and the Japanese had not found it. I have often thought that it was the Peking man in that locker," Logan said.

A number of marines mentioned William Foley and Herman Davis. Sometimes it was only a passing reference since they knew Davis (with Beeman) had initiated discussions of Peking man at the marine reunions, but others saw Foley and Davis as keys to the mystery.

"The Chinese friends of Dr. Foley seem to be a good place to pick up the trail of the lost containers," one marine said.

It was well known that Foley had many civilian friends in all parts of China, and that he had seen the fossils in the PUMC laboratory and was aware of their significance. He was a highly skilled, knowledgeable physician who carried skulls with him for teaching purposes. Surviving marines were insistent that Foley, and his longtime assistant Davis, would have a lot to say about the fossils, perhaps more than anyone still alive.

If the sum total of the surviving marines' stories means anything, it is that the fossils could have traveled a number of routes and ended up in a variety of hands. Many of the marines admitted knowing or hearing about the fossils in the days before and after they were taken prisoner, but few stated positively that they had been personally aware of the fossils' location. All had suspicions about the Japanese; some claimed the invaders were thorough and meticulous, others believed they were clumsy, incredibly ignorant soldiers capable of throwing anything away.

Not one of the marines admitted having seen the fossils. Not one of them saw an open locker containing anything resembling fossil remains, or witnessed what could have been the transferral of fossils from one container to another. Yet not a single surviving marine discounts the possibility that the fossils might have been taken from Camp Holcomb by an American marine or one of his connections.

PART TWO

To China

CHAPTER **5**

A Phone Call in Chicago

It began with a telephone call. It was early April 1972, and the caller was Mai Lai Wei, a Chinese-American woman from New York who had traveled with me on one of the several trips I had organized as president of the Greek Heritage Foundation.° Mai Lai was elated over President Nixon's trip to China six weeks earlier and felt it was the first step toward normalization of diplomatic relations with the Chinese government. She was calling to suggest that the Foundation organize a China tour.

Mai Lai and her husband had fled from China in 1949 when the Chiang Kai-shek regime fell. They left everything and everyone there, and were eager to return and visit friends and relatives. Mai Lai talked somewhat sympathetically about the Chinese government. "People are being fed and clothed now," she said, "and regardless of whether or not you believe in communism, that is the beginning of the rebuilding of China."

° The Greek Heritage Foundation is a non-profit organization devoted to the advancement of world art and culture, and to the study of the application of "cultural values of ancient Hellenic heritage to the contemporary world." It organizes annual cultural symposiums, grants scholarships, funds student exchange programs, and helps promote peace and understanding among peoples of the world.

She had called specifically to urge me to apply for a visa. I liked the idea; I didn't know very much about China or its culture but I was sure it would be an exciting place to visit. But I was also aware that no American tourists had yet visited the People's Republic of China and the prospect of such a trip appeared dim. Mai Lai was insistent; she told me that I should apply immediately.

"Just believe me," she said. "I can't go into all the reasons now, but take my advice and apply for a visa through the Chinese Embassy in Ottawa." We talked of other things for a few more minutes before she hung up. I remembered it as a pleasant conversation, but in the days that followed I didn't follow up on it.

About ten days later, Mai Lai called again and asked me if I had made the application. I confessed that I had not. Again, she urged me to take her advice and said, somewhat cryptically, "You must apply; you may be surprised."

I took the second conversation more seriously. There had to be a reason for Mai Lai's persistence, so I made up my mind to go through with the application procedure. I asked Valerie Valentine, a close friend and longtime assistant in my business affairs, to approach the Chinese Embassy in Ottawa. I also decided to propose the idea of a trip to China to the directors of the Harvard Club, the alumni organization in which I've always taken an active part. The club might apply for visas and perhaps coordinate the effort through members who worked in the State Department.

I had done business on the international level before, and I knew that something more than wishful thinking and a few influential friends were needed to get me into a country that had been tightly sealed for more than twenty years. I knew I would have to appeal directly to Chairman Mao himself.

More than 400,000 visa applications were on file in the Chinese Embassy in Ottawa, and I wanted mine and those of my traveling mates to stick out of the pile like a flag waving in Mao's face from the midst of a vast crowd.

I went to a local university bookstore and bought a copy of *Quotations from Chairman Mao*, the "little red book." I knew it was the bible of the People's Republic, that it was read and revered by millions. I decided to use it as my own text, and after hours of study I began to grasp the style and logic of Mao's thought. I had already sent four cables to Mao and Chou En-lai requesting that my visa application be accepted and had received only silence in return. I would try one more cable and then I would feel I had done everything possible.

Chairman Mao Tse-tung
Hangchow
People's Republic of China

The snows have come and gone, and a thousand flowers have blossomed since our friends have asked through your embassy in Ottawa to be welcomed to your great country. As you have written, we have been made to feel like a blind man catching sparrows. We ask why four cables and a telephone call bring only a great wall of silence. Another revered sage, like yourself, has said courtesy is the sister of friendship. In the name of courtesy, and friendship, may we respectfully ask you to intercede on behalf of the Greek Heritage Foundation and the Harvard Club group who have been made to believe they may depart for a cultural visit to your country on May 14.

So far, no visas have been authorized. If Edgar Snow were alive, he would befriend us and Henry Kissinger will speak for us. Professor Willis Barnstone has translated your strong poetry so beautifully that it has been honored with Book-of-the-Month Club Selection. Is a scholar and translator of your poetry not welcome with our group? A sad no is better than a long silence and a happy yes is a cultural step forward for peace and friendship for our two countries. We respectfully ask and hope for the courtesy of a prompt and favorable reply. All good wishes.

Christopher G. Janus

The cable cost me $124. I will never be sure what effect it had but within two weeks I received word from Ottawa that my visa application had been approved.

Once the notice came, things happened fast. I was told that the visa permitted me to take along four friends, fellow members of the Harvard Club or anyone else, or I could go alone. I was asked if I could leave that week. I was more than happy to hurry. Valerie Valentine flew to Ottawa to get our visas. Reservations were made immediately for flights to Seattle, Tokyo, and Hong Kong, and on May 31, just seven weeks after Mai Lai Wei had first suggested it, we were on our way to China.

In many ways we were an unusual group. The first cultural visitors to enter China after President Nixon's visit consisted of a businesswoman, two stockbrokers, a college professor, and a lawyer. All were professionals, all were over thirty, all were well-educated, veteran world travelers, and all were able, at a moment's notice, to drop whatever they were doing and set off for three weeks in a country they had never visited before. We were a group that could derive a host of impressions from the visit and be able to relate them to Americans on a variety of levels. We were, by the same token, a group that could tell the Chinese a lot about America.

Valerie Valentine is a thirtyish, round-faced woman with blonde hair and a ready smile. She is a director of the Greek Heritage Foundation and is the executive administrator of that organization. She is equally at home handling brokerage accounts or organizing a nationwide backgammon tournament. She has a genius for keeping dates and names and places in order, an asset to any tour.

Frank Voysey is an investment banker, a wealthy businessman whose voice hints of his British background. Tall and distinguished looking, he dresses impeccably and sports a well-trimmed mustache. But for all his propriety and demeanor, Voysey was chosen for his easygoing nature. People as amiable as Valerie and Frank relieve the inevitable tension of a long journey.

Everett Hollis is a sixty-two-year-old lawyer and friend. His casual dress and manner contrast with the fastidiousness of Frank Voysey, but his intelligence, his affability, and his talent for asking interesting, enlightening questions added much to the group.

Willis Barnstone, the fourth member of the party, was my ace in the hole. Barnstone is a professor of comparative literature at Indiana University and had published the translation of Chairman Mao's poetry mentioned in my cable. I wanted to present Mao with a specially bound copy of the poems and see if Barnstone could appeal to the chairman and possibly Chou En-lai, also a poet, on an artistic, nonpolitical level.

Barnstone had impressed me not only because he was an excellent writer and gifted poet, but because he was deeply involved in evaluating modern China.

As for myself, I have done many things in my sixty-two years—I've traveled widely and been involved in several different professions—but most of my experience has been in international business. I've sold companies in Europe to others in South America. I've traded goods and materials when international currencies weren't available—machine tools to Turkey for figs and pistachio nuts, clothing to Greece for olive oil. In 1948, I had arranged for the sale of a shipment of ball bearings but the buyer was short of funds. I asked him if he had anything to barter, and he said that he had a Mercedes limousine that had been the private touring car of Adolf Hitler. I took a chance, made the trade and brought the limousine to America. Despite much initial criticism, I was able to exhibit the car throughout the country and use the proceeds to aid Greek war orphans.

I am comfortable with such endeavors, but my education and my background have taught me that there are many things in life more important than money, and perhaps that is why I have derived such pleasure from the Greek Heritage Foundation, its travels, and its various cultural activities. I've also enjoyed the efforts I've put into *Poetry* magazine, the Harvard Club, and a host of other clubs and organizations in Chicago. I have never shied away from the unusual or the unfamiliar. Risk taking, I guess, is second-nature with me.

All the same, as I sat on that Hong Kong-bound plane, I still wasn't sure if I was the man to take on the Chinese. I asked myself, what am I doing going to China? I'm not an expert; I haven't studied the culture. Indeed, until a few short weeks ago I hadn't even entertained the thought of visiting China. Suddenly I felt very tired. I simply wanted to sleep, not to think. But the mysterious circumstances surrounding my sudden rush to China lingered in the back of my mind, and the mystery was to deepen when the plane landed in Hong Kong.

None of us was a stranger to Hong Kong. We had all been

there many times. We went directly to the beautiful Peninsula Hotel, whose reception area is world famous. It always reminds me of old Shepheard's in Cairo. The group decided to rest from the long plane ride and shake off the effects of the time change—it's thirteen hours later in Hong Kong than in New York. I was restless and wandered to the bar. I sat down and ordered a glass of orange juice. Soon, the tall Chinese sitting next to me drinking stingers began a casual conversation. He spoke English perfectly.

He asked me what I was doing in Hong Kong and when I told him he was very interested; he hadn't met many Americans going to China, he said, and he knew how difficult it was for Americans to get visas. He mentioned, I believe, that he was in the manufacturing business and understood the difficulties involved in foreign travel. Then he asked if there was anything he could do for us while we were in Hong Kong.

The strangest thing about him was his familiar manner. He acted as though he knew me, and was soon talking to me as if he were an old friend. He certainly found out a lot more about me than I of him. He said his name was Nine, and when I remarked that it seemed an odd name, he replied it was just a family name. I let it go at that.

We had a couple of days before boarding the train for Canton (Kwangchow), our first stop in China. So we decided to spend a day or two sightseeing in Macao, the small Portuguese island that lies a few miles out from Hong Kong. We had not made reservations; we simply turned up at the docks the following day. There we were confronted by a line of hundreds of people waiting to board the hydrofoil for the island. With so many people ahead of us, we decided the trip was hopeless, and turned back to the hotel. Suddenly, we were approached by the smiling Mr. Nine.

"Hello. Are you going to Macao?" he inquired.

"Well, we had hoped to, but it doesn't look like it's possible," I said.

"Don't worry about that," he said with a grin. "Let me help you."

He left us for a few minutes and returned holding a sheaf of tickets. Without explanation he gave me the tickets and hurried us aboard. Before leaving us, Mr. Nine recommended a hotel on Macao and wished us a pleasant time.

The boat trip took about an hour. When we reached the island we sought out the hotel Mr. Nine had mentioned and a respite from the intense heat. To our surprise, we were greeted in the lobby by the smiling Mr. Nine. He laughed and explained that he had some business on the island and wasn't it a coincidence to meet again. He then asked us what we planned to do. When he learned that we were interested in sightseeing, he suggested a car-rental agency. We took his advice again, and enjoyed a very pleasant and informative day.

That night I was eager to try my luck at Macao's famous gambling casinos. I like roulette, an easy and relaxing game whose 35 to 1 odds hold an irresistible appeal for me. For about a half hour I casually played the numbers and enjoyed myself even though I was losing. Then I was joined by Mr. Nine.

Nine asked me how I was doing and I told him I had lost a few hundred dollars but that I was having fun. In his by now familiar gracious but confident way, he suggested I ask the croupier what number to play. "Sometimes they're luckier than you are," he said.

The croupier told me to play number seventeen. I did, and lost. I played it a second time and again lost. I was betting five dollars on each number and after the second loss I glanced at Nine. He insisted I keep playing seventeen.

On the third try, the number came up. I collected my winnings and played it again. Seventeen hit once more. Then the croupier told me to play number twelve and it immediately came up. Soon, I had won back my losses and made a profit of several hundred dollars. I decided to stop there. I thanked Mr. Nine and retired for the night.

Throughout the rest of our stay on Macao, I was conscious of Nine's presence. I can't say for certain that we were shadowed by him or anyone else, but I did always sense there was someone nearby, a figure, unidentifiable and perhaps a figment of my imagination.

The next day, we boarded the boat with little difficulty. Nine was also on board. We greeted him warmly, but made no further plans to meet.

Back at the Peninsula Hotel, I asked a bell captain to recommend a restaurant. He directed us to one which featured entertainment and dancing. Shortly after we had been seated, we spotted the ubiquitous Mr. Nine dancing with an attractive Chinese woman.

The most curious meeting with Nine turned out to be our last one. Three days after our arrival in Hong Kong we departed by train for China. Nine was at the railroad station to see us off, just as he had been at the harbor two days earlier. Before we left he said something peculiar. "I hope you find the treasure," were his words; I recall them quite clearly.

I asked him what he meant, but all he said was, "Never mind. I hope you have a great trip." As usual, he was smiling broadly. Then he said, "I hope you bring home the bacon," or some similar Western expression. Again, I was puzzled but found no answer in his smile.

The train pulled out, leaving Mr. Nine smiling on the platform. We were not to see or hear from him again.

China overwhelmed us: We were awed by the country's beauty, its history, its expanse. And we were fascinated by the way of life we observed. We awakened each morning to the hiss of thousands of bicycles skimming over the pavement, taking the Chinese to work. We watched masses of workers, led by exercise instructors, taking time each morning to keep physically fit. Except for the children in their colorful, almost gaudy outfits, the people seemed drab and unattractive. Men and women alike were dressed in loose blouses and baggy trousers. The clothes were utilitarian and unisex, and the impression was one of uniformity. No one had greater privileges, or a larger apartment, or a higher social status because of his profession. Unlike the Soviets, who have been known to make exceptions for favored citizens, the Chinese seem untouched by such capitalistic values.

I wore a pair of loafers, slacks, and a shirt, and usually a tie. As our group's representative I tried to maintain a certain formality. The others, because of the extreme heat, were less formal. I bought a lightweight cotton "Mao jacket" in a Cantonese department store. The so-called people's jacket, which sold for about five dollars, was a comfortable alternative to a Western suit jacket. Blue and gray wool jackets sold for about thirty or forty dollars, and I bought two or three of them as well.

I had taken along a collapsible cane that could be carried

in a small bag. It was somewhat like a magician's cane and opened with a sharp tap. This turned out to be a wonderful attention-getter. The Chinese children were delighted by it and would run, scream, and giggle when I demonstrated how it opened and closed. They were happy and healthy-looking children, full of natural curiosity.

In Canton, we took pictures of modern China against the background of its centuries-old culture. We saw the marble dragon head reliefs; we walked through tombs and gardens; we learned about the Ming and Tang dynasties. Everywhere we heard the words "revolutionary," "self-reliance," and "struggle." Modern Chinese culture is a political culture: ballet, opera, acrobatics, art, and ping-pong all exist within and for the state and the party.

In a Canton art institute, we saw a collection of three or four hundred paintings, in each of which Chairman Mao was pictured. Disturbed by this overwhelming display of art as propaganda, I asked the young Chinese woman who was our interpreter, "Don't you think that art is created for its own sake?"

She quickly answered, "No, art is to serve the people. Everything is to serve the people."

Was nothing, then, a private experience, an entity or concept that existed on its own terms? I wondered about love. Had the state and party appropriated that too? I asked her about her husband whom, I presumed, she had married for the universal reason of love.

She paused for a moment but then answered, "No, I did it for the people."

From Canton we traveled to Peking, where we stayed at the Nationalities Hotel. It is a new structure, somewhat gloomy and functional, reserved for foreigners and their guides. Although the accommodations were simple, the Chinese went out of their way to be hospitable to us. There was no air conditioning, but each room was equipped with elaborate electric fans which, by means of a button-dotted console, pushed air in all directions and at a variety of speeds. Two or three thermoses filled with hot and cold tea were in constant supply, one of the luxuries afforded us to counteract the intense heat. We bathed three and four times a day in an effort to keep cool, and after a while, our most familiar posture while in our rooms was reclining in the clean, cold tubs.

Another climatic feature is the dry yellow dust that sweeps over the country. It was necessary to keep our hotel windows tightly shut at all times, but one afternoon I inadvertently left a window open during my nap and awoke to find the room covered with a thick layer of yellow dust.

Though windows and doors were kept closed against the elements, we left our rooms unlocked without concern. We were continually impressed by the atmosphere of safety and respect for our privacy. There was no hint of surveillance or shadowing, as is often the case in the Soviet Union. Willis Barnstone often left the group at the end of the day's activities and wandered alone about the city and country-side. Once he left his cameras on a city street; they were promptly returned to him. A Chinese official even returned a pair of undershorts he had left in a hotel.

We had been in Peking about ten days when our Foreign Ministry representative knocked on the door of my room. He politely inquired if we were enjoying our stay. It was a question I had not stopped to ask myself. I was a visitor, with no political or diplomatic credentials, and the fact that I was even inside the country still seemed remarkable. But the man had asked the question as a preface to the announcement that we would see something very important the next day. He did not explain and said only that we should be ready by eight the next morning. After the usual courtesies, he left.

I was thrilled and immediately began to wonder if that something meant our desired meeting with Chou En-lai.

The next morning we were picked up by three Shanghais, the efficient little Chinese-made auto that looks much like a 1947 Plymouth, and proceeded toward the village of Choukoutien, which lies about thirty miles outside of Peking. Shortly after we crossed a bridge outside Peking, we were stopped by uniformed guards who were adamant that our little caravan could not pass. The Choukoutien area was apparently restricted and our guides took great pains to convince the guards that we should be allowed to proceed. After a heated argument we traveled on and finally arrived in Choukoutien.

It was an unspectacular place, hilly with few houses and little to see. Moreover, the day was dusty, incredibly hot, and hardly suited to sightseeing. Our caravan headed for the outskirts of the village, toward a low range of hills, and stopped in front of a single-story, L-shaped building.

We were told that we had arrived at the Peking Man Museum, which had been closed to the public for several years. We were met in the road by a lean, sallow man dressed in the usual people's jacket. He was Wu Ching-chih, the director of the museum, and he was to guide us through it. He led us inside and, true to the custom we had encountered everywhere, we were taken to a small room where the Chinese sat across from us, served tea, and discussed what we were about to see.

Wu knew some English, but he chose to speak through our interpreters to explain what Peking man was all about. He gave us a brief history of the discovery, asked if we had any questions, then led us into the museum. For two hours we viewed charts, pictures, and replicas of the fossils in the museum's small galleries. Then Wu took us into the Upper and Lower Caves directly behind the building. The caves appeared to be untouched by any efforts at renovation for the benefit of tourists. The deep, rock-strewn tunnels were about ten feet high and fifteen feet wide and angled fifty or so feet into the mountain. Inside the caves, Wu pointed out areas where key fossil discoveries had been made.

Back in the museum building, we studied the life-like sculpture in the center of the main gallery. It is a slouching ape-man, his calves and thighs straining under the weight of a small, deer-like animal draped over his shoulder. The figure has almost no neck and stares ahead with eyes deeply set beneath a thick, bony ridge along his eyebrows. His head is flat and small, the chin tucked under his collar bone.

We gazed at the figure and listened while Wu recited names and dates and facts pertinent to the creature. We were all impressed and questioned Wu about Peking man even though Frank Voysey was the only one of us who possessed more than a laymen's knowledge of physical anthropology. As time passed, the heat began to get to me and I became somewhat bored. Finally I turned to Wu and asked him whether we could see the fossils. Wu looked at me and for a moment his genial nature changed.

"But this is why we have brought you here," Wu said. "Surely you know the story."

Beckoning for an interpreter, Wu took me into an adjoining room. There he told me the story of the fossils and the anthropological significance of their discovery. He emphasized that the fossils were considered a treasured part of the heritage of the Chinese people. Then he paused and looked at me incredulously.

"But don't you know what happened?" he asked.

I admitted that I did not. He continued with the story of Claire Taschdjian and Franz Weidenreich, of the train to Chinwangtao, the Japanese invasion, and the fossils' disappearance. He mentioned various rumors and bits of scattered information filled with discrepancies he could not explain.

Wu paused once again, then said, "We have a contract with you to return the fossils."

I was abashed and replied quickly, "No, Doctor, you have no contract with me."

"But we do," he insisted. "We have a contract with the United States Marines to return our fossils."

He took my arm and looked at me intensely. "Will you help us find these fossils? Will you return them to us?"

I was startled by the request. Throughout our visit, the Chinese had never asked us for anything. They had been very careful, either out of courtesy or policy, not to make us feel obligated to do anything. Yet this man was pleading with me for help, in a most extraordinary endeavor, apparently believing that somehow I could help return the fossils to China.

I couldn't imagine why Wu had seized upon me as the

79

questor of this national treasure. I had an uncle who had found and excavated the site of Plato's Academy in Greece but surely I hadn't been mistaken for that relative. Was it simply because I was an American, the first Wu had run across? That seemed as unlikely. It certainly wasn't a joke. I had observed enough about China to know the Chinese don't joke and there was no denying the urgency in Wu's plea. I thought of Henry Kissinger's efforts to once again open the door to China and didn't want to slam it shut right there in the Peking Man Museum. It seemed reasonable that I could make a few calls and some casual inquiries when I returned to the United States, and that, I thought, would discharge my duty. In any event, I was becoming uncomfortable with the pleading eyes of Dr. Wu upon me.

"Of course I'll help," I said.

Although my encounter with Wu Ching-chih unsettled me, his request did not evoke fantasies of treasure hunts, mysterious strangers, and clandestine encounters. On the trip back to Peking, I discussed the disappearance of the fossils with Valerie Valentine and Frank Voysey, but in a casual way. Frank has made a hobby of anthropology and was aware of the discovery and importance of Peking man, but he confessed that he had not been aware of the fossils' disappearance. He was inclined to believe that the Chinese were not deadly serious in their request that I find the fossils, and that our visit to the museum was simply a part of the standard itinerary. But I was intrigued.

"Wouldn't it be great, Frank, if we could find them?" I asked.

We talked briefly about offering a small sum as a reward,

then the subject changed and Peking man was, for the moment, forgotten.

Valerie, however, immediately realized that Wu's challenge had appealed to me. We discussed trying to get some publicity once we returned home, perhaps a story in a national magazine, and I could see that her mind was already making arrangements.

In the hours that followed our trip to Choukoutien, I remembered a talk I had had with Henry Kissinger when I met with him and his aides shortly before leaving for China. He gave me some idea of what to expect from the Chinese and hinted at what our group might encounter.

"Whatever you do, Chris," he told me, "don't pinch the girls."

I thought he was joking, but he was serious. He told me about some Englishmen who had gone to China and ended up in jail because of their improprieties with Chinese women. China is a girl-watcher's wasteland, Dr. Kissinger implied. Their mores are different, and nothing even casually suggestive is appropriate.

Dr. Kissinger also warned me of the delicate nature of the relationship between China and the United States and said that American visitors should be very sensitive and careful not to offend or annoy the Chinese.

"I don't have to tell you how to act abroad," he said to me. "But we haven't officially recognized China yet and we have to be diplomatic and polite. *If there is anything you can do for them, anything that will increase the good will between the two countries, you should obviously take pains to do it.*"

Thinking back on what Dr. Kissinger had said, I wondered if it had been a coincidence or if he had had some prior knowledge of what the Chinese were going to ask me. Whatever the truth might be, it was becoming clear to me that I had to follow Dr. Kissinger's directive and do whatever I could to help the Chinese. Our search for Peking man was now a reality.

CHAPTER **6**
Peking Duck Party

My newfound interest in Peking man did not escape the notice of the Chinese. From the moment we left the museum, the mood of our hosts—the translators and guides and the government officials—changed dramatically. Up to that point, everything had been courteous and formal. They always addressed me as "the responsible person"—the Chinese term for the leader of a group—in a humorless way. There was never any levity, never any casual banter. But shortly after we returned from Choukoutien, our hosts announced that they were going to give us a party. They asked what kind of food we'd like and I suggested Peking duck.

"Of course," beamed one of the guides. "That is exactly what we have planned for you." (The Chinese were always careful to ask us what we wanted to do or see or eat, but they always proceeded according to their own, seemingly prearranged, plans. Inquiries were a matter of form. Perhaps this is simply because tourism is a new phenomenon in China and small changes in plan can create great difficulties.) "We will take you to the Peking Duck Restaurant, and we will have a party and enjoy ourselves."

"We are going to enjoy ourselves," he had said. It was the

first time I had ever heard any Chinese mention the word enjoy.

At 7:30, the Chinese arrived at the hotel in three Shanghais and drove us to the restaurant. There were the five of us, plus translators and the usual government ministers—about twelve people in all. The restaurant was charming. It reminded me of an old-fashioned family restaurant, and had a warm, comfortable atmosphere. We were not to dine in the main dining room, but were led to a small, private room, in the center of which was a round table set with simple flatware and a bouquet of flowers.

Before we sat down, Valerie Valentine went around to each place and set out some favors she had picked up at a department store earlier in the day. She felt that we should show our appreciation to our hosts and give them some small gift as a token of our gratitude. The Chinese were somewhat uncomfortable about this. Though they were more relaxed than they had ever been, they were still cautious and were reluctant to accept gratuities. They did take the favors, however, and seemed pleased.

By then it was time to have an opening drink at the small side table. We were served the traditional Mao Tai, an incredibly strong alcoholic drink the Chinese enjoy immensely. I generally don't like alcohol but I handled this rather well, which made a great impression on the Chinese. They laughed and patted me on the back.

The sudden conviviality struck me. This was not an official, formal reception, rather it was a party in a private, very intimate dining room. There was a fireplace, a sofa, a floral bouquet. It was totally out of character with the China we had seen, but much in keeping with the China I had known prior to the revolution. Back then, a diplomat would take his guests to this kind of a room and enjoy relaxed conversation; an ambassador might even take his mistress to such a place. But in Mao's China things are not private and cozy; they are open and political. And that evening our Chinese hosts, trained though they were in the manners of modern China, seemed to know that the old-fashioned, intimate atmosphere could be found in this room. It occurred to me that this change in tone, which followed our trip to the Peking Man Museum, was obviously not a coincidence.

As we were enjoying the Mao Tai, Valerie decided she would make another attempt to relax our relationship with our hosts. Back in Chicago, she had said that if it ever became possible for us to throw a party for the Chinese, we would have to have something uniquely American to add. She had brought along a bottle of Scotch and a tape recorder with tapes of jazz, rock, and even some Greek music. She suggested we listen to some of the tapes although she was not sure if this would make a bad impression or somehow anger the Chinese. They did not object, which I observed to be yet another turn in our relationship.

We finally sat down at the table. I was eager to see how they prepared Peking duck in the Peking Duck Restaurant, in Peking, and I was not disappointed. Three waiters dressed in immaculate white uniforms marched in, each carrying a large dish with a whole roast duck on it. It looked the way a roast duck should look, with vegetables and oranges and other garnishes arranged around it. They ceremoniously displayed the platters and then took it all back to the kitchen where the meat was chopped up in very small pieces. They returned with three plates stacked with small bits of duck and placed them in the center of the table. We proceeded to devour the food.

After the meal, we played more of Valerie's tapes. Willis Barnstone asked Valerie to dance and they treated us all to a lively display of contemporary American culture.

Valerie caused a bit of a stir, as she had ever since we arrived in China. Her striking good looks and stylish dress are in marked contrast to the no-nonsense attire of the Chinese women. On that particular evening she was wearing a rather elegant cocktail dress, as befitted the occasion, in American eyes at least. It was a gay and frivolous interlude and, although our hosts did not join the dancing, several of the women later told Valerie that they had enjoyed the music. In some respects, our American entertainment was a flop. Perhaps our hosts were afraid of what their superiors would think, or perhaps they genuinely did not enjoy it.

The conversation, however, was open and informal and often cluttered as everyone tried to speak at once. At last the Foreign Ministry representative got up and offered a toast to friendship between the People's Republic of China and America. He said he hoped it was the first of a number of reciprocal visits.

I answered the toast with one of my own, recalling a thought Dr. Kissinger had impressed to me. "Friendship," I said, "is something that is received only when it is given away."

Then one of the translators rose and expressed her appreciation for the little gifts we had given them and said she expressly wanted to thank the beautiful Miss Valentine. This may seem like a mere courtesy, but contrasted with the formal, almost rigid relationship that had prevailed until that night, those words were remarkably personal.

The ministry representative offered one last toast. Turning to me, he said, "To you, Mr. Janus, for coming to China and sharing our hospitality; and for agreeing to bring back to China the fossils of Peking man."

A few days later, the Foreign Ministry, the government group in charge of our stay, gave a tea in our honor. The reception took place in a large, open meeting room, quite unlike the intimate surroundings at the Peking Duck Restaurant. In the center of the room was a table covered with sweets, cakes, and nuts, and waiters walked around serving a variety of drinks. A photographer mingled with the crowd, snapping many informal shots.

There were many other Chinese at this tea, members of the Foreign Ministry and the Office of Tourism, interpreters and security people. Many of them had accompanied us at different times on our visit. We sat in a semicircle; an official of the ministry sat to my right, a tourist official to my left, interpreters flanked the two men, and the rest of our group was seated down the line. It was again formal and proper; the customary official toasts were exchanged.

In my remarks I proposed a student exchange program between our countries, as well as a visit by the Chicago Symphony Orchestra. Both projects were of great interest to me and I talked about them in detail with the top officials. I was also hoping to hear that forty-five visa applications for the Greek Heritage Foundation/Harvard Club cultural group had been approved. The Chinese seemed receptive to all this and I was hopeful that my extra push would bring results.

87

Later others got up and proposed toasts and made informal speeches. Everett Hollis stood up and talked about law and China's legal system. Frank Voysey briefly mentioned the Stock Exchange and how it might work under the Chinese system. Other topics were casually discussed, and the time passed smoothly. Then more pictures were taken, and the Chinese stood up to bring the tea to a close.

As we were preparing to leave, one of the ministry officials turned to me, and with an interpreter, led me to one corner of the room. It seemed to me quite conspicuous, not at all unlike the manner in which Dr. Wu had taken me aside in the Peking Man Museum. And, in fact, the subject of our conversation was to be the same.

"Tell me," he asked, "how did you enjoy your visit to the Peking Man Museum?"

I said the museum was fascinating. I was beginning to realize how important these fossils really were, I continued, and even though I did not know their whole story, I was committed to help recover them in every way possible.

His next statement startled me. "You know, Mr. Janus, we are very appreciative of your interest. But I want you to know that this is of special interest to Chairman Mao. Whatever you do will be most appreciated."

The official went on to say that he had seen some of the correspondence that had preceded our visit. He had apparently read the cable I had addressed to Chairman Mao. He smiled and said he thought the cable was very amusing and very clever. Throughout our little talk, the official seemed well informed on almost all aspects of our trip and our reason for being in China.

As he spoke, it was quite obvious to me that specific plans had been made by the Chinese before we arrived. There was an element of cunning in the casual way in which they had brought us to the Peking Man Museum and dropped the task of recovering the fossils in my lap.

Shortly before we left Peking, we watched a parade given for a visiting delegation from Chile. The Chinese are accomplished parade-givers and think little of assembling as many as a million people along a parade route. The celebration for the Chileans was no exception. As we looked on, a ministry official turned to me and said emphatically, "Return the fossils to us, and you will be a national hero in China. The parade we will give you, Mr. Janus, will be ten times as big as this!"

It was not a casual remark, nor was it said in jest. It was an expression of the change in who and what I was that had occurred in the eyes of the Chinese. A pattern of events was becoming clear to me: the moment Wu Ching-chih challenged me with the idea that the Americans had a responsibility to return the fossils, the time when the ministry official informed me of Mao's concern, and now this talk of national heroes and parades. For some reason I had been singled out among thousands of potential Americans to renew the search for Peking man. Nothing had been coincidental. I began to realize that the Chinese can be very calculating; there is a purpose behind everything they do.

Seemingly, the Chinese had picked me because they had decided that I was rich enough, or had enough connections, or was intelligent enough, or perhaps foolhardy enough to try to find Peking man. And, as with most things on our tour, they were right.

How I would go about my search for Peking man was still vague in my mind when we reembarked for Hong Kong, but my plans were to materialize suddenly when we were met by a group of reporters at the border. I was immediately asked about Peking man. What about the search? What was I going to do? Was I positive I could find the fossils?

I had little time to ponder who had tipped off the reporters. In the face of that barrage of questions, I heard myself announce a $5,000 reward for information leading to the recovery of the fossils. Within hours the story was on the front pages of several newspapers. I had put money where my mouth was, and I was officially searching for Peking man.

CHAPTER **7**
Two Strong Leads

The $5,000 reward kicked off my campaign to rekindle interest in the fossils. The next step was to generate a storm of publicity. I wanted the story of my search to circulate throughout the world but, because of the information I already had about the fossils' disappearance, I wanted to focus upon Taiwan, Japan, and the United States. The reception I received from the press in Hong Kong on June 16 assured that the search would be given worldwide notice. The reward offer was the story the reporters wanted, and they followed up the impromptu press session with in-depth interviews and front-page stories in the Hong Kong papers.

When we arrived in Tokyo three days later, I repeated my offer of reward to a receptive press corps which reported the story throughout that country. At a dinner given in our honor by the Associated Press, reporters peppered me with questions about Peking man. The AP had picked up the momentum from the Hong Kong press and soon its wire service was buzzing with the story.

A week later, I arrived in Chicago and began sifting through early responses to the reward offer. I was still not sure if what was happening had been programmed by the

Chinese from the very beginning—from the gracious Mr. Nine to the surprisingly knowledgeable Hong Kong reporters—but I was too involved by this time to let these questions bother me.

Most of the more than 300 letters, telegrams, and phone calls were quite worthless, the predictible impedimenta of any treasure hunt. The authors of the letters usually revealed themselves as jokers. In the first paragraph of a carefully written four-page letter, one woman identified herself as the illegitimate child of Charles Lindbergh and claimed she could not only establish her heritage but knew where the fossils were. Some letters carefully postulated an answer to the mystery; others seized upon random facts and drew ridiculous conclusions. One letter suggested that I abandon the search and give the reward money in exchange for the letter-writer's wife.

In the middle of this avalanche, I got a call from a woman who refused to reveal her identity. She called me at the Harvard Club, my usual New York haunt but one which I had not made public. I later found out that she had traced me through Chuck Chamberlain, an Associated Press reporter in Chicago who had done a piece on the fossils. The woman was referred to Chamberlain by the AP office in New York. He in turn directed her to my Chicago office and my secretary there told her how to contact me in New York.

The woman called me at about 9:30 A.M. She asked for me by name and identified herself merely as a friend of Edgar Snow. I replied that I had had tea with the late journalist's wife, Lois, during our stopover in Hong Kong. We spoke for a few minutes about the family. I knew Lois only slightly and my caller admitted that she had been better acquainted with Snow's first wife, but that brief exchange established a rapport between us. After scores of phony calls I was relieved to talk with someone who seemed legitimate and serious. The brief reference to the well-known sinophile and his family seemed to establish the woman's credentials.

She went on haltingly, as though straining for the right words. "You know I have . . . I am the widow of a marine. My husband died seven years ago. I was quite young when I married him . . . and I think I have something, in view of this story I've been reading, that might be of interest to you."

"Great, great, come over and let's talk about it," I said quickly. "Can we meet for lunch?"

There was a long silence as she considered my invitation. Unable to see her face, I could only speculate about what was happening at the other end of the line, but I sensed her tension and, perhaps, fear.

"No, no . . . you don't understand," she finally said. "You know, my husband was a marine and he told me about the fossils a long time ago before he died. And he indicated—he never told me—but he indicated that he got the fossils through considerable violence. He said that at least one person lost his life in connection with the fossils. I don't know what he meant by that because that is all he would say.

"My husband was a very honest man, he was a direct man, and he was a poor man. But he was hard-working and very brave. I don't know whether he did anything wrong by taking these fossils or how he got them. 'Now remember,' he told me, 'this is stolen property. Maybe I'm entitled to it, maybe you're entitled to it as much as anybody.' Then he told me the fossils were his legacy to me.

"There was a footlocker up in the attic for years. It was sealed and I didn't look into it or ask him about it, but I was aware it was there. A few months before he died, he told me about the fossils. He said they were the famous Peking man fossils. And some day, he said, if not already, they would be worth a great deal of money.

"Mr. Janus, I am not a woman of any means and I don't know what my legal rights are, but I have the fossils and I'm sure they're what you're looking for. I just wanted to tell you and see what we can do about it."

She stopped and waited for my response. I had taken the call lying in bed, but now I was sitting on its edge, fully awake.

"Can you come here? I mean, are you in New York? Can we have lunch?" I pressed.

"I don't know about that," she said. "My husband warned me about this. He said that if anyone found out I had the fossils, it would be very dangerous.

"Quite frankly, I don't understand all the intrigue. But he told me the story only a few months before he died of cancer. He said to be very careful. He was always very secretive about this."

I tried to persuade her to come, but she said no, she wanted to be very careful. She said she'd seen my picture in a New Jersey newspaper and knew my search had been well publicized. She didn't want to be seen with me for fear that someone would make the connection. Anyone aware of her involvement who knew that she was the widow of a marine would put two and two together and her life would be in danger, she maintained.

I tried to assure her that I had no motives other than simply locating the fossils. I didn't even care about seeing them myself. I wanted to find them, pay whatever was necessary, and have them returned to the People's Republic of China, where they belonged.

"Well," she said, "I do want to talk to you. I want to believe that you are sincere."

I kept pressing her to meet with me, but my caller wasn't persuaded and said she needed time to think about it. I told her I intended to stay in New York for a few more days before returning to Chicago.

"Why don't we just meet today," I said. I am seldom so insistent, but this unknown woman could vanish with the click of the receiver. I didn't want to lose her, my first solid lead.

"All right," she finally said. "We'll have to think of a place. I want to be very careful and you mustn't think I feel persecuted or anything, because I don't. I'm just trying to take precautions because these things are worth a lot of money and a lot of people are looking for them. I think I have them and I don't want to get into a lot of trouble."

I offered a suggestion. "Do you want to pick up a taxi and meet me in front of the club? I could get in and we'll drive around."

She was silent.

"Or you could come to the club. I'll meet you in the back room, or my room, or whatever."

"Oh, no, no, I wouldn't want to do that," she broke in.

But she had an idea of her own: the Empire State Building.

"I'll go up to the observatory—I think it's on the eighty-sixth or eighty-seventh floor—and we can at least meet. You can see me and hear my story again and know that what I'm telling you is the truth."

"It isn't very private up there," I countered.

"I know," she replied. "That's why I've chosen it."

We agreed to meet there at 12:30 that afternoon.

"I'll find you," she assured me. "I know what you look like."

As I hung up, I had the feeling that I should have insisted on meeting her at the Harvard Club or a similarly private place. The Empire State Building struck me as ridiculous. But the cinematic quality of the suggested rendezvous fit in with much of what I had already encountered in my search; there was no reason why the intrigue should cease.

I wandered about the observation deck for about ten minutes. I looked inquiringly at several women standing nearby only to be greeted by cold glances from husbands and boyfriends. The woman had said she'd be able to recognize me from newspaper photos she had seen. Was she a crank; was she perhaps on the deck now, enjoying a rather bad joke? I ground out my cigarette and considered leaving when suddenly she was at my side. She was a tall woman, about five-foot-eight, with black hair. Perhaps in her late thirties, she was moderately well-dressed, but wore a three-quarter-length coat which appeared too warm for that day in June. She addressed me haltingly.

"How do you do? You know I called you this morning."

"Hello, I'm Christopher Janus," I responded, expecting that she would introduce herself.

But she only answered, "I know."

She glanced away casually, scanning the observation deck. It wasn't crowded. Clumps of tourists milled around, some pausing to gaze through the coin-operated telescopes that line the edge of the open-air deck.

Cave at Choukoutien, near Peking, where the Peking man fossils were found in 1926. *Photo courtesy of Field Museum of Natural History, Chicago.*

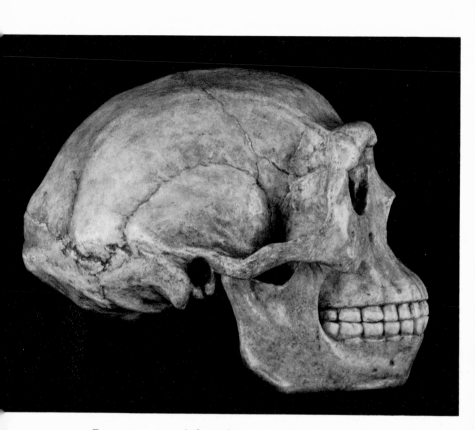

Reconstruction of the Peking man skull at the Museum of Natural History. *Photo courtesy of the American Museum of Natural History, New York.*

Claire Taschdjian, the last person to see the Peking man fossils when, at the Peking Union Medical College, she packed them for shipment.

Tientsin Marine Detachment at the barracks in Tientsin, China, lined up for the commanding officer's inspection, 1941. This is where the North China marines were assembled by the Japanese after the outbreak of war.

Five marine pharmacist mates (now called hospital corpsmen), left to right: Bill Hunt, Ed "Pappy" Fox, Herman Davis, Earl Johnson, and Danny Wolmer; the dog was named Ergo. This picture was taken in late fall 1941.

Herman Davis' superb physical condition was probably one of the reasons he was able to endure and survive five years in Japanese prison camps. He saved his own life and those of many others.

The barracks at Camp Holcomb; it was from a window about thirty feet beyond the right corner of the picture that Davis aimed his machine gun. The gun was mounted on footlockers packed with the fossils.

Colonel Ashurst approaching the Japanese Embassy to surrender, December 8, 1941.

Peking detachment, North China Marines Embassy Guard, being marched to prisoner of war camp, January 1942. This picture was taken by a marine at the risk of his life.

Peking detachment, North China marines boarding train to prison camp at Tientsin. This was shortly after capture on December 8, 1941.

Pictured at reception given the Greek Heritage Foundation group, Peking, June 1972, are, left to right: Yi Shao-jung, Wang Kuan-sen, Yeh Chuen-chien, Willis Barnstone, Yeuh Dai-hong, Chia Ai-mei, Valerie Valentine, Everett L. Hollis, Christopher G. Janus, Yang Kun-shu, Frank E. Voysey, Hsih Chen-huan, Liu Ju-kun, Liu. The Chinese interpreters and guides did not usually make the entire trip with their visitors; at each city the group was greeted by new guides.

Valerie Valentine with the visas for the People's Republic of China. Ottawa, May 1972.

Members of the Greek Heritage Foundation at cave where fossils were unearthed, June 1972. *Photo courtesy of Field Museum of Natural History, Chicago.*

Wu Ching-chih, director of Peking Man Museum in Choukoutien.

Display of primitive tools used by Peking man. Peking Man Museum, Choukoutien, People's Republic of China, June 1972.

The Peking man reconstruction by Harry Shapiro. *Photo courtesy of the American Museum of Natural History, New York.*

A picture taken by tourist Max Niealai, of Düsseldorff, Germany. This picture was sent to Janus after Niealai heard of the search for the fossils and the strange meeting on the top of the Empire State Building. This picture shows Janus with the Empire State Building woman.

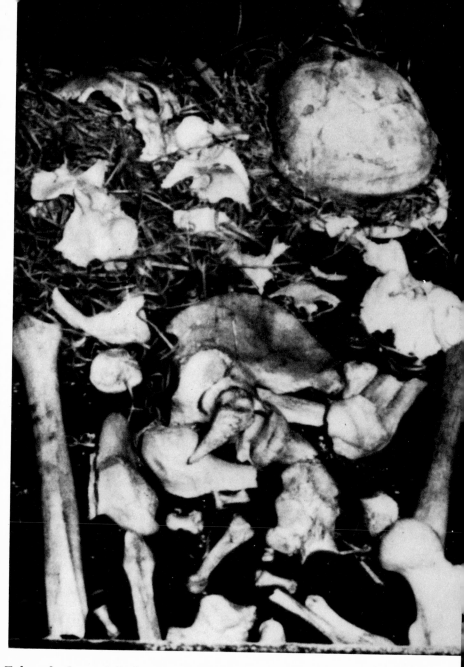

Enlarged photo of "Peking man fossils" given to Janus by the
Empire State Building woman.

Christopher Janus flanked by two friends in Taiwan. Ho Hao-tien, on the right, is director of the National History Museum, Taipei, Taiwan. It was Ho Hao-tien who had suggested following the trail to Thailand. *Photo courtesy of Niki Janus.*

Janus in front of a painting of Peking man, Taiwan, August 1973, during the second trip to Asia.

Janus and Dr. Chiang Fu-tsung, who is director of the National Palace Museum, Taiwan.

Left to right: Dr. Li Chi, president of the China Academy, Taiwan, and a renowned archeologist, and Dr. Chien Hsaeh-liang, its present director. Chris Janus stands behind the two experts.

Christopher Janus with part of the Peking man exhibit at the National History Museum in Taiwan. By coincidence, this exhibit was being prepared during Janus' visit.

"You are an impatient man, Mr. Janus. I must insist that you let me handle this my way. I have my reasons. The safety of my client is of primary importance and we're not taking any chances. You can take all the chances you want, Mr. Janus, that's your business.

"For the time being, let's just say that there are a lot of people interested in my client's fossils and they're valuable. I've read about your offer of five thousand dollars for information, but that's not what we're talking about. We are talking about five hundred thousand dollars, and I wonder where you or the Greek Heritage Foundation is going to get this money. Is it going to come from our government? Or will it come from the Chinese? The fossils are worth every cent of it, but where is it going to come from?"

"Mr. Seng, I'll let you handle things your way if you let me handle the question of money my way. Where the funds come from is an academic question. What I'm interested in first is authentication of the fossils."

"The first step, before anything is examined, is some assurance that the rights and the safety of my client will be protected."

"All right. What do you want?"

"There are several things. First, I want some sort of ruling from the district attorney's office in New York stating that my client will not be prosecuted and the fossils will not be confiscated. I want that kind of a letter, if not from the district attorney then from the federal government."

"All right, Mr. Seng. Is it Seng? S–e–n–g?"

"Harrison Seng, yes."

"That's an interesting name. How did the Harrison get in there?"

"You ask too many questions. Stick to the subject. My identity and qualifications are of no real importance."

"They're important to me. I want to know that I'm dealing with a responsible person." I was beginning to have my doubts. Seng's language befitted a fast-talking promoter more than the lawyer he claimed to be.

"We are not playing games with you," he insisted. "We have no interest in this thing beyond receiving fair payment for the return of the fossils. With all the agony that my client has gone through, with all the risks that her husband took, the money we're asking is a very small amount."

"All right, I'll get you the letter. I'll get some sort of statement that I hope will satisfy you. Where shall I send it?"

"For the time being, I'll call you. How long do you think it will take?"

"I don't know. I could have someone phone you from the State Department or the FBI . . ."

"Let's not get the FBI into this. They have no authority in this case."

"All right, maybe I can get the district attorney in New York to phone . . ."

"No, no, phoning is not appropriate. I want you to get an official letter saying my client will not be prosecuted. Period. After you get that and we're satisfied, we'll talk about the second phase."

"What's the second phase?"

"The arrangements for the money."

"Look, Mr. Seng, you and your client don't get anything until you've proven you have the fossils and they're identified as Peking man."

"All right, you get me the letter and we'll make

arrangements to have somebody authenticate them. But meanwhile, when I call you again you had better have a pretty complete plan for how the money is going to be turned over to us. I'll call you, shall we say in a couple of weeks?"

"Fine, but I wish you'd let me know where I can reach you."

"That won't be necessary."

I thought about the conversation for some time. It provided a new insight into the Empire State Building woman and one I had to admit I didn't like. I had gotten a very bad impression of the lawyer—if he indeed *was* a lawyer—on the basis of our talk. I didn't want to misjudge him, but his manner was hostile and devious. He diminished my confidence in his client, and I began to wonder if I was being conned.

My optimism was revived by news from a different source. During one of my trips to New York I had discussed the mystery of the fossils with Dr. Harry Shapiro, curator emeritus of the Department of Anthropology at the American Museum of Natural History in New York. Dr. Shapiro had been personally involved with Peking man since its discovery. He had visited Choukoutien in the thirties and observed Franz Weidenreich's research in the Cenozoic Laboratory. In articles and books, Shapiro had chronicled the events following the disappearance of the fossils and was a leading authority on the mystery.

He was one of the experts to whom I sent the Empire State Building woman's snapshot. Dr. Shapiro was impressed. He said that it was a bad photo and that it showed many bones that clearly were not part of the Peking man collection, but the skull pictured in the upper right-hand corner of the locker appeared to be of the right proportions for Peking man. It was "interesting," Dr. Shapiro said, even though he could not confirm that the skull was authentic.

✳ Dr. Shapiro's observations were supported by Dr. Glen Cole of the Field Museum of Natural History in Chicago. Each man, however, insisted that he would have to examine the fossils in person before making any definite judgement.

Their remarks were encouraging. I felt a kind of guarded optimism which would not permit me to discount the Empire State Building woman. All the same, I could not overlook the fact that the two experts had cited only one part of the pictured collection—the skull in the upper right-hand corner of the trunk—as possibly being authentic. Since more than 175 specimens had comprised the PUMC collection, a considerable number of the fossils might still be hidden elsewhere.

I had yet to investigate a third major source of information: Dr. William Foley and Herman Davis. Dr. Foley is currently a highly respected cardiovascular surgeon in New York City; Herman Davis is employed as his office assistant. I was anxious to talk to both men and hear their stories myself.

Davis received me enthusiastically. He greeted me in his white intern's smock at Dr. Foley's offices on East Sixty-eighth Street. We went into Davis' small private office where he showed me photographs of the North China marines.

Davis identified the pictures and listed the names of his fellow soldiers. It was obvious that he was proud of the group and the fact that he had been a part of it. He told me which of the marines had died within the past seven years, and tried to deduce which one might have left a widow with a collection of fossils. I could give him little help. I had neither a name nor a description of the woman's late husband. We knew only that he had been a marine, but Davis and I agreed that he might not have been one of the marines at Chinwangtao or even in North China.

Davis mentioned the possibility that the woman's husband might have come across the fossils once they had left Camp Holcomb if, indeed, they ever did. He suggested I tell Dr. Foley about the woman; perhaps something would ring a bell for him. Foley, however, was not available at the time, so Davis proceeded to explain what he could about Camp Holcomb and the Japanese invasion. He drew a diagram of the camp (see page 41) which indicated where the railroad siding, the rifle range, the barracks, etc. had been located. He described precisely how and where he took the footlockers off the train and transported them to his room. He marked an X on the diagram to show where he had waited behind the machine gun on the morning of December 8, 1941.

He was very cordial and informative, ready to answer whatever questions I had. He mentioned the North China marines' yearly reunions and suggested I attend one. Finally, he recommended that I tell my story to Dr. Foley.

The next day I met Foley for lunch. In the midst of a bustling hospital cafeteria, Foley commended me for the way I was conducting the search.

"At last someone is going about this in a systematic way," he said. "I think your reward offer is a step in the right direction."

Yet Foley was not as open or cordial as Davis had been. He is a short, distinguished-looking man in his sixties who wears half-glasses for reading and looks over them in a rigid, almost chilling manner. He is very serious, not at all quick to smile and he talked with me in a brisk, businesslike tone. He was, of course, a busy man. I had learned about Foley's ill-treatment as a prisoner of war and perceived that he was still bitter about it. Foley agreed that it was a good idea to interview the remaining marines and said, to my delight, that he would help me in any way he could.

The story of the Empire State Building woman fascinated him and he said he thought it was definitely worth pursuing. He said he had no idea how the woman's husband might have gotten hold of a footlocker, but he did not discount the possibility. Foley was reticent about the details of his own footlockers. He continually referred me to articles he had written; since he had told his story before, he said, he didn't care to review the particulars. He was polite but firm about it. I had the distinct impression that he would be offended if I pressed him with more questioning. But it was equally clear to me that sooner or later my questions would have to be asked, and answered.

CHAPTER **9**
Enter the Nationalists

Late that summer—August 1972—I wrote a letter to Henry
Kissinger describing the situation that had developed between me and Harrison Seng. I had been unable to reach
Dr. Kissinger by telephone because he was out of the
country. One of his aides suggested that I put in writing
exactly what I wanted from the State Department.

About two weeks later, Harrison Seng phoned again. He
asked if I had heard anything from the State Department or
the New York District Attorney's office. I told him I had
not, but that I expected a letter soon.

"Fine, fine. That's all I wanted to know," Seng said. "My
client is going out of the country for three weeks but that
won't interfere with our schedule. Shall I call you again in
about ten days?"

I told him ten days would be fine and then we launched
into what seemed like the hundredth reprise of an old,
familiar song: I asked about authentication, he countered
with protection of his client's interests and safety; he
inquired about my source of funds, I insisted on the
irrelevance of his question. I paused and tried a new
subject.

"Tell me, was your client's husband one of the marines in North China?"

"You ask too many questions, Mr. Janus," was the brusque reply. "I don't want you snooping around trying to find out who my client is. That will just complicate things. I've told you that a lot of other people are as interested in these fossils as you are. It isn't quite as simple as you make it out to be. I'll call you again in ten days."

But ten days passed and Seng did not call. In the meantime, I contacted John Holdridge of the National Security Council. Dr. Kissinger was consumed by his efforts to work out an end to the Vietnam War. Holdridge provided an interim reply from the State Department regarding the status of the fossils and said follow-up work by the legal staff of the department was in progress. He was able to assure everyone involved that the United States government had no intention of prosecuting anyone connected with the fossils. When Seng did call some weeks later, I was able to read him a letter from John Richardson, Jr., Assistant Secretary for Educational and Cultural Affairs in the State Department, which elaborately confirmed the government's wish not to prosecute the Empire State Building woman or anyone else involved. The letter read as follows:

December 21, 1972

Dear Mr. Janus,

I am following up the interim reply of John Holdridge of the National Security Council to your letter of September 2, 1972 to Dr. Henry Kissinger about the Peking man fossils.

The legal staff of the Department of State has looked into the question of potential legal liability for the possessor of the fossils, should they turn out to be genuine.

On the basis of the facts you have set out, the Federal

Government would not be interested in prosecution against the "New York woman" mentioned in your letter. Moreover, it seems likely that the applicable Statute of Limitations has expired and thus the question of possible criminal proceedings, for example, under 18 U.S.C. 2314 and 2315 and 541 and 542 (transportation of stolen goods and false statements) would not arise.

Should the lady in question wish further to explore this aspect of the problem, she might be best advised to consult a private attorney. While the formal granting of immunity is possible under law, there are other less difficult and equally effective ways of seeking and securing assurances in appropriate cases, as for example, through the action of the relevant United States Attorney.

I see absolutely no possibility of a financial reward from the United States Government for the return of the fossils. I hope therefore that you will be able to persuade the lady in question that she will be doing not only her duty, but a genuine service to the United States, in coming forward with the fossils to make them available for return to the People's Republic of China.

The Department of State believes, as you do, that if these fossils are in the United States, they should be returned to Peking, and that their return would be an important contribution toward improved relations between the United States and the People's Republic of China. Thus, the Department of State is deeply interested in the recovery and return of these fossils and can assure you that the Federal Government has no interest, on the facts presented, in having any criminal action pursued.

If you have further questions, please do not hesitate to write me. You may also wish to contact the Director of our Office of East Asian and Pacific Programs in the Bureau of Educational and Cultural Affairs, Mr. Francis B. Tenny.

I hope this information will be of some help to you in ascertaining whether the fossils are genuine, and if they are, in bringing about their return to China.

Sincerely yours,

(signed)   John Richardson, Jr.
Assistant Secretary for
Educational and Cultural
Affairs

I read the letter slowly and Seng often asked me to repeat certain parts. It was apparent that he was taking notes, if not writing down the complete text of the letter. He was full of questions about details; some were answered in the letter, others were not.

"Why don't you let me send you a copy of the letter," I said.

"No, no. That won't be necessary. I'll contact you when I want a copy."

"Well, is the letter satisfactory?"

"It covered all the points except one. It says the federal government is not interested in prosecuting my client, the statute of limitations had run out, and so on. What the letter does not say is that the fossils won't be confiscated."

"I have no interest in confiscating them." I was beginning to sound exasperated.

"But what about the government?"

Seng was adamant about that point and I finally agreed to see if I could get some kind of statement on it from Washington. He would call again in two weeks.

While the dealings with the elusive Mr. Seng were in progress, Andrew Sze became even more hostile toward my efforts. He was offended by the amount of publicity the search was drawing; ironically, the same publicity had initially brought Sze forward with his secret. He recoiled at it all and said articles and notices in magazines and newspapers were detrimental to the search because they would scare off any Chinese who had information to give. Sze maintained that the Chinese do not enjoy the glare of publicity, they are quiet, unspectacular people who prefer to move cautiously. And, Sze added, he wanted to proceed this way himself.

He became more and more irritated whenever I contacted him by phone. He said he was certain now that his Chinese friends would not cooperate and risk exposure. He told me that he had sent a letter in late November to his friend in Taiwan, but had received no reply. It was a sure sign, he said, that the Chinese would not cooperate until things quieted down and they felt it was safe to make disclosures.

I urged Sze to tell me the names of his Taiwan friends so that I or someone else could contact them.

"Look, I'm not going to tell you anything anymore," Sze said. "You have someone from the State Department contact me and I will see what can be done. But I do not want to deal with you anymore."

Sze sounded as if he were on the edge of hysteria. He threatened to sue me if I did not stop bothering him. I reminded him that his name had never been used in any of the articles on the hunt and that no one but me knew his connection with the fossils. It was to no avail—Sze would not be calmed. My long-distance conversations between Chicago and New York often ended with the Chinese gentleman loudly slamming down the receiver.

I finally decided to release his name to State Department officials and see if they could make progress with him. Sze had approached me with very solid information; he had not wanted any money, and seemed to possess a stable, responsible background. But he had never relaxed the barrier of fear and mistrust he had erected between us. He

123

reiterated the danger involved in the hunt though he never explicitly told me what the danger was.

I wanted to know if Sze was conning me. My instincts told me otherwise, but the stalling tactics he had employed of late made me wonder. Nothing had followed a logical course since I had begun the hunt. And though I had many reasons to believe Sze and his story, I pressed myself to remain skeptical. The time had come to see if others could make any significant inroads.

As fall of 1972 turned into winter of 1973, very little progress was made. Officials in the State Department assured me that the new demand by Harrison Seng about confiscation of the fossils would be answered. It was March 2 before a written reply came from Francis B. Tenny, Director of the Office of East Asian and Pacific Programs.

Dear Mr. Janus,

You ask in your letter of February 14 "what is to prevent the Government from confiscating the Peking man fossils once they are turned over to someone for authentication or once they are sold to the Greek Heritage Foundation?"

As Mr. Richardson wrote you on December 21, the Department considers that return of the fossils to Peking would serve the national interest. Confiscation is not a way of doing this. Indeed, I am advised, having regard to our prior correspondence, that there is no adequate basis at this time warranting action by the United States government of the sort you ask about. Naturally, the United States would not take the fossils without legal right. Were the presence of the fossils in the United States established, however, the government would wish to afford them any appropriate protection within its power.

Your deep interest in locating and returning these fossils to the Chinese people is greatly appreciated.

Sincerely yours,
(signed)   Francis B. Tenny

Harrison Seng contacted me shortly after I had recieved Tenny's letter. He called at a Manhattan restaurant where I was having lunch with friends. I was by that time accustomed to such interruptions but his ingenious pursuit continued to amaze me. He wanted to know if I had received the letter and when I said I had but didn't have it with me, he asked me simply to summarize its contents. He was, of course, interested in the matter of confiscation. I was able to assure him on that point. I could have predicted his next words.

"Okay, now let's talk about the money."

"You have a one-track mind about the money, Mr. Seng, but I have a one-track mind about authentication."

"The one goes with the other," Seng said. "How long do you think it will take to authenticate the fossils once we turn them over to you? Can it be done in a very short time?"

"I don't know how long it would take, Mr. Seng. I would guess at least a week. Some of these tests are rather sophisticated."

"Forget it," Seng retorted. "This whole transaction has to be made in a very short period of time."

"What do you mean?"

"A couple of hours. The funds must be brought on the same afternoon that the tests are going to be made. Once you're satisfied that they are authentic, you get the fossils, we get the money."

"You know, Mr. Seng, this is beginning to sound more and more like a ransom."

I tried to appeal to the lawyer in him. "Look, we'll get the money. I'll put it in escrow with a letter of credit payable on receipt of a letter stating that the fossils are genuine. You can take that letter to the bank and get the money."

"Janus, I've told you, we can't do it that way. The money has to be in cash. You bring it to a meeting place we'll agree upon. You can bring whatever experts you want, I don't care who they are. Just make sure they are experts and don't try to bring along a lot of other people, if you know what I mean."

"I will have to discuss this with a number of authorities on the fossils to see if it can be done."

"Well, if it can't be done this way, it can't be done at all."

I was quite sure Seng was serious. If the meeting were not set up exactly as he wished, things might not come off at all, and the whole deal would fall through. I attempted another approach.

"Listen, forget about bringing *all* the fossils. Just bring the skull. Or if not the skull, bring a few other specimens, perhaps some teeth or something representative of what you have. On that basis, we can take all the time we need and not jeopardize the entire collection."

"Nothing doing. We do it all or nothing. You talk with your experts. Set things up so the whole examination can be done, say, between two and five o'clock in the afternoon. Can you make those arrangements in a couple of days?"

"I don't know, Seng. I'm here on other business and I don't know how long it will take to get the scientists together. I don't know if such an examination can be done like this. Why don't you give me a week and maybe by that time I'll have an answer. Where can I expect you to call me?"

"Don't worry about that. I'll find you. You just do your end of the thing and we'll get along fine."

"Okay, Seng. Let me ask you one more question. Do you by any chance know Andrew Sze?"

"Who?"

"Andrew Sze."

"No, I don't think so. Who is he?"

"Never mind. You're sure you don't know him?"

"Quite sure."

"Do you know Dr. William Foley?"

Seng paused. "Dr. Foley? He's one of the marines, of course. He was in your newspaper story, wasn't he?"

"Yes, he was in the story, but I'm asking if you or your client know him."

"I don't know. I don't think so. I don't know that my client knows him personally, but of course, she has read about him."

"Are you sure?"

"I'm not sure, but I don't really think so."

"Okay, Mr. Seng. I'll get to what we talked about and expect to hear from you in a week or so."

The demands of the Empire State Building woman and her lawyer, and the fearful, evasive Andrew Sze totally dominated my thoughts in the weeks that followed. I tried to link personalities and motives, attempted to analyze everything I had learned in the hope I had not missed an obvious connection. The trail had widened and it now pointed in a number of different directions; but several paths could lead to the same door. I am neither a professional nor an amateur detective but I forced myself to think along clear, logical lines, keeping an open mind to any new lead, yet zeroing in on what I already knew.

This meant I had to look as far back into the mystery as I could. A lot of things had happened both in China and the United States, things that at the time may not have seemed significant but which, upon reflection, suddenly fell into place. I cautioned myself against overreacting, against grasping at straws. However, some of my friends who were aware of my *sleuthing accused me of doing so. The tendency was almost unavoidable.

My focus on past events, particularly those that might explain why I was picked by the Chinese to find Peking man, brought to light an incident I had all but forgotten. It was another startling connection.

While we were still in Peking, we met a large delegation from the Philippines. The group consisted almost entirely of women, and since they seemed to have the same itinerary as we did, we often encountered them. They were a university group, ostensibly in China to learn more of Mao's philosophy. We talked amiably, comparing experiences and sharing details of our backgrounds. One of the leaders of the Philippine delegation invited Willis Barnstone to Manila to lecture (an invitation he later accepted), and she also suggested that I come, too, even though I told her I really had nothing to lecture about.

This was the basis of a casual relationship. But one young woman of about twenty-three appeared to take a special interest in me. She was quite attentive to everything I had to say and often questioned me and the others about what we were doing, how we happened to be in China, where we were from, and similar inquiries. I didn't pay much attention to her inquisitiveness, and simply treated her in a friendly manner.

A few days after we arrived in Hangchow the women from the Philippines followed. One night, I heard a knock at my hotel-room door and opened it to find the young woman. I was immediately troubled by her presence, careful as I was to avoid giving the Chinese the wrong impressions. I remembered Dr. Kissinger's warning about the Chinese and women and I knew that our hosts might misunderstand even the most harmless of relationships. Earlier in the trip, Frank Voysey had good-naturedly touched a female interpreter on the shoulder and he was immediately scolded by a Chinese and told to take his hand off her.

But she greeted me in her usual friendly way and gave the impression that this was simply a social visit. I was not so sure, and after some casual small talk, I tried to find out why she had come.

She laughed off my inquiries and tried to continue the chit-chat. She was smoking a cigarette and gave me one. And then she changed course.

"You know something about those fossils?" she asked.

"What fossils?"

"You know, the Peking man fossils you saw when you went to the museum."

"Yes, but how do you know about them?"

"Well, I know a lot of things."

"About Peking man?"

"After the war many Chinese fled the country when Chiang Kai-shek left in nineteen forty-nine. A lot of them went to Taiwan with him and a lot of them escaped to Hong Kong. But a lot of the refugees went to Manila. So you may want to go there because many people think that maybe the fossils are in Manila."

"I've not heard that," I replied. "I've heard of the possibility that they were in Taiwan, Tokyo, Hong Kong, or maybe still in China, maybe even in New York. But I've not heard Manila mentioned before."

129

"Oh yes, they might be in Manila," she said. "You should come."

A few minutes after that she got up, just as abruptly as she had come, and headed for the door.

"You come to Manila," she said.

Then she thanked me for the tea we had, and left.

I forgot about this meeting soon after that. It was a small event, and I had many other things on my mind. The woman and I had talked seriously about some things, joked about others, and perhaps she had made pointed references that went completely past me; I wasn't sure. But as I recalled the meeting later, something about it stuck in my mind.

It was her name. I suddenly remembered her name. It was Maria Sze.

During this period, I devised other means of getting information from people who would not want to talk directly with me. My research on the North China marines and the statements of Andrew Sze pointed me toward Taiwan and the Nationalist government. A number of people had insisted the fossils had made their way there with Chiang Kai-shek, along with thousands of other valuable articles of all descriptions. That view was largely unsubstantiated, and often expressed by individuals with a political and even emotional axe to grind. But the importance of Taiwan in the mystery began to loom larger in my mind.

I found, however, that my dealings with officials in the Nationalist Chinese Embassy in Washington seldom proceeded past superficial courtesy. I suspected that the information they provided me was not all they had, and they might be more forthcoming with someone not known to be involved in the search. It was clear that the overwhelming political friction between the Nationalist and Communist Chinese governments tended to obscure the most basic information.

I decided to use a different approach. I contacted a few friends who are high-school teachers and asked them if they would have some of their students make inquiries about Peking man. I discussed the matter with several teachers in Chicago, one in Avon, Connecticut, and another on the West Coast, and they agreed to assign the project in their classes.

About fifty letters were sent out to Peking, Taipei, and to the various embassies in this country. They were simple inquiries stating that the writer had read about the search in his local newspaper and was seeking further information. Some asked to be told the complete story, others asked about specific details, still others asked whether the recipient knew where the fossils were and to whom he believed the fossils belonged.

Shortly thereafter, the replies began to come in. Most were perfunctory responses thanking the students for their interest in the fossils but stating that that particular office had no information on the subject. But one student in Connecticut received a startling response from the Nationalist Chinese Embassy in Washington, D.C. On embassy stationery, press counselor Y.T. Chen told George Marecki of Farmington, Connecticut, that his government was in control of the Peking man situation.

His letter read as follows:

Dear Mr. Marecki,

Many thanks for your letter, inquiring about information on the

"Peking man." However, we would like to inform you that the "Peking man" was found by our government—the Republic of China—in 1929 at Choukoutien near Peiping. It was moved to Taiwan, one of the Chinese provinces, along with the Government of the Republic of China in 1949, when the Chinese Communists overran the mainland under the instigation and support of the Soviet Union. As a matter of fact, not only the "Peking man" but even all those things on the Chinese mainland still belong to the Republic of China, rather than the Peiping regime, which has not been established on the consent of the Chinese people, as evidenced by millions of refugees escaped to Hong Kong and some other places.

Sincerely,

(signed)   Y.T. Chen

For the first time a Taiwan official admitted having any knowledge of the fossils and, most significantly, stated unequivocably that the fossils were on the island, if not specifically in the possession of the government. I was dubious so I contacted William Braden, a reporter for the Chicago *Sun-Times*, and asked him to confront the Nationalist Chinese Embassy with the letter. If its information were true, the search would be over.

Braden called Chen and congratulated him on solving the Peking man mystery. But the official quickly retreated. He said that a mistake had been made, that the fossils were not in Taiwan. Though he insisted the fossils did belong to the Nationalist Chinese, he said he knew only that they had been lost many years ago. When Braden confronted him with the letter to George Marecki, Chen insisted that a mistake had been made.

Copies of the letter were sent to officials in Taiwan, but there were no replies. Instead, they sent a press release from the Chinese Information Service about the efforts of Dr. Li

Chi, a seventy-seven-year-old archeologist at the National Taiwan University in Taipei. The release chronicled Dr. Li's thirty-year effort to recover Peking man and stated that he had run into the same dead-end as everyone else. The release also mentioned me and the reward for the return of the fossils. But the essential message was that nobody knew where Peking man was. It made no mention of the smug assurances of Y.T. Chen that the fossils were safe in the hands of their "rightful owners."

The brief encounter with Mr. Chen whetted my appetite for a trip to Taiwan. I had an uncanny sense that the Nationalist Chinese, from Chen to the mysterious friends of Andrew Sze, knew more about Peking man than they wanted the rest of the world to find out.

PART THREE

In the National Interest

CHAPTER **10**
The FBI and the CIA

In June 1972, a few days after I returned from China, I received a call from the FBI. The agency was anxious to gather all the information possible about modern-day China and they hoped I would agree to a "casual visit" and tell them about my trip. I consented to see them, but I had reservations.

I was convinced that one of the main reasons I and the others were allowed into China was because we were not a government group. I preferred it that way. Other trips, the ones I had conducted through the Greek Heritage Foundation, had always been arranged purely as cultural tours with the emphasis on person-to-person contact between visitor and host. The mark of government or "official" status had always been carefully avoided.

I realized, of course, that a tour of China at this point in history was a different matter, and communications from the FBI and the Central Intelligence Agency prior to my departure had reminded me of that fact. Still, I was adamant about keeping the trip unofficial. I did not want to act as an envoy or quasi-agent for any American intelligence agency. The Chinese had accepted me as a friendly visitor with cultural interests in China and that is how I would go.

I was not hostile to the pretrip approaches of the FBI and the CIA. I simply declined their offers of help—without even asking what they had in mind. My only preparation consisted of the conferences I had in Washington with Dr. Kissinger and other members of the State Department and later with the Consul General in Hong Kong, with whom I discussed such things as social amenities. I simply wanted to know what to expect and sought advice on how to react when confronting a nation that had not talked to the United States, other than in diplomatic jargon, for more than twenty-five years.

The FBI agents talked with me in my office for about an hour. Their questions were those I would have expected of inquiring school teachers. What was acupuncture like? Did you see Chairman Mao? What did the Chinese show you? How did they act toward the group? I filled them in on our itinerary and added as many pertinent details as I could remember. It was information I would have given anyone. Later in the conversation it was my turn to ask the agents questions which had bothered me for weeks.

"I met a strange man, a Chinese, in Hong Kong by the name of Mr. Nine, who seemed to be one step ahead of me at every turn. I thought he might be a Communist agent but perhaps he was one of your men."

The agents were fascinated with the tale of Nine, but they denied knowing anything about him and said they were positive Nine was not an FBI agent. They said they would look into it. As I retold the story, I remembered that before the trip I had been assured by Roy K. Moore, then head of the FBI's Chicago office, that the agency would help with anything—customs, reservations, transit visas—if I and the others needed assistance. I had declined the offer, and forgotten it but the ubiquitous Mr. Nine brought it back to mind.

The FBI visit was followed by a call from the CIA. Their approach was a little different, but I was resolute about treating their inquiries in the same manner. The CIA agents brought an agency report on China and offered to let me see it. Their questions were related to information gathered in the report. What did I think about the progress of China's industrialization? Did the people seem happy? What about political prisoners? Had I seen or heard anything of them? How did the police and the military act? What was I permitted or prohibited to do?

Though the questions differed, my replies contained essentially the same information I had given the FBI. I described the freedom the group had enjoyed and recited at length my thoughts about a visit I had made one night to another American staying in a Peking hotel three or four miles away from ours. The point of the story, in my mind, was the apparent ease with which I moved about the city. At no time during the visit did I or the others have the impression that we could not do what we wanted. It was not a touchy matter and describing it to the CIA did not seem to me to taint the trusting relationship we had maintained with the Chinese.

The initial interviews passed routinely. I compared notes with the other members of the group—all of whom had been interviewed—and learned that their sessions had been quite similar. Some weeks passed before I heard from the FBI again. By then, I was completely engrossed in the search for Peking man.

This time the FBI said that their interest was more than a routine matter; they had received a directive from Washington to assist me in any way possible. The agent was vague about the reasons behind the FBI's involvement, but he emphasized that the search for the fossils had been declared an activity "in the national interest."

This phrase sounded to me like a slogan which could have meant just about anything. But I was told that such a classification can mobilize the various federal investigative and intelligence agencies. In the words of the agent, "national interest" meant "How can we help you?"

Though I had resisted government involvement up to that time, I now had second thoughts. I felt that my involvement in the search originated because I was a private citizen, not a government figure. Though the Chinese may have had it all planned from the beginning, they had still presented the challenge to me in a personal, unofficial way, through the uncompromising eyes of Dr. Wu, the museum director. If I were to involve the United States government in the person of its intelligence agencies, sooner or later the matter would go through similar Chinese governmental agencies. Then it would be out of my hands, as a private citizen, and become a political issue. On the other hand, whether I accepted the FBI's help or not, I knew they would remain keenly attuned to my activities. The search had in any case ceased to be a purely private matter.

I decided to let the FBI help me, particularly in gathering data from people I could not easily interview. After I had begun the search, I was contacted by other people who at one time or another had been involved in the hunt or were still at it. Some of them were overseas, some were in other parts of the United States. I asked agents to interview these people to consolidate the available information and perhaps turn up something I had missed.

It was only after the initial weeks of the search had lapsed into months, and my leads began to play themselves out, that I decided to increase FBI involvement. It was by then spring 1973, and I had made contact with the most important figures in the search. I asked FBI agents to interview some of the surviving North China marines who had been involved in transporting the fossils. And after some hesitation, I also asked them to interview Herman Davis and Dr. William Foley. I reasoned that these people might be persuaded to divulge to federal agents information that they may have held back from me. I added Claire Taschdjian to the list and finally, though reluctantly, Andrew Sze.

Sze's hostility toward me, and his request to talk to a representative of the federal government, left me with few options. My action was also spurred by a coincidence. During a meeting with an FBI agent in my office, I was interrupted by a long-distance telephone call from Sze. I announced this to the agent and asked him to listen in on the conversation so he could get some impression of Sze.

Sze began where he had previously left off. He accused me of betraying him with outrageous publicity that drew too much attention to the search. He repeated that he no longer wished to see me, that he had nothing more to say to me.

"Listen, Mr. Sze," I said. "You are a citizen of the United States, is that right?"

"Yes," Sze replied. "My wife and I are citizens and very proud to be."

"Well, I want you to know that the State Department has declared it in the national interest that we find the fossils. So I would say that it is your patriotic duty, perhaps even your obligation, to help us out. If you have a friend in Taiwan or wherever who knows where the fossils are, then you owe it to your country to reveal this person's name.

"Just give me the name and I will go to Taiwan and see him. We've been through this before, I know that, but I wanted to urge you, *in the national interest.*"

"Well," Sze answered, "if it is so much in the national interest, you have an official of the government call me."

That gave me my opening.

"Okay, Sze, you've got a deal. A government official will get in touch with you and I want you to tell him everything. You can tell him what you haven't told me."

The stage was conveniently set. I asked the agent if he would pursue the matter. He nodded assent, but did not appear enthusiastic.

"Is there a problem here? You said you wanted to cooperate in every way. . . ."

The agent hesitated, then replied. "We don't know, but we have a feeling that interviewing Andrew Sze could pose a problem for the State Department. It could involve investigation in Taiwan. You must remember that we still recognize the Nationalist government there. The point is that the State Department does not want to get into a conflict between Taiwan and the People's Republic over these fossils."

But I was insistent.

"Look, it's important to me. It's important for the whole search that I find out if Andrew Sze is lying or what. I'm trying to find out why a man like Sze would go to all the trouble he has. He doesn't want publicity, he doesn't want money. It leads me to believe that he must have something."

The agent did not argue and promised me that a special agent would be assigned to talk to Sze.

With that the conversation was closed. I was unsure about what I had started by opening the door to the FBI. I could only continue my own investigation and hope the agents would complement it.

Two months later, on June 8, 1973, an agent called me and said the first of the bureau's reports had been filed. He said the report was for the benefit of the State Department and was not to be seen by unauthorized civilians—meaning me—but the agent in charge agreed to meet with me.

We talked in O'Connell's, a small restaurant in the same building as my office. The agent had a seventeen-page summary of the report with him. He explained that it was only the first report and that interviews were far from being completed.

"But unfortunately," the agent said, "I can't let you have it."

I frowned. "What's the point of doing it if you're not going to let me have it?"

"Well, I can let you read it. But technically, the bureau has done this for the State Department and it is up to them to decide what they're going to do with it."

I felt ill at ease with the report in my hands. It had "FBI" stamped all over it and I was not sure how important or how vital it would be to my search. To peruse it casually over coffee in a crowded restaurant did not seem proper. After some pressuring the agent finally agreed to let me read it in the privacy of my office.

The report began with a brief outline of the vital facts of the search as I had explained them. The next section was a lengthy summation of an interview with Dr. Foley in which

Foley repeated much the same story he had told me. He detailed his duties in China and the circumstances of the surrender of the North China marines. He said he had only seen the fossils at the Peking Union Medical College laboratories and that he was aware that they had been delivered to Chinwangtao after they left Peking. Foley also reiterated his belief in the integrity of Colonel Ashurst and said he believed the colonel "most certainly" would have delivered the fossils to the American Museum of Natural History had he been successful in transporting them to the United States.

The report also suggested that Foley might be an integral part of the mystery. In a section in which Foley spoke of his present contacts with Chinese friends, the report read:

Foley said that he has been in contact with an individual in the United States who is in turn in contact with the people in Tientsin in whose custody these footlockers were placed. He said, however, that he could not divulge the name of the person in the United States or the names of the people in China as they would be in great danger if their identities became known. He said that he feels in the future when the political climate is better, he may divulge the names of these people to determine if they still have the footlockers he gave them and if they still are intact and determine what they contain.

Foley would go no further than that.

The report also summarized an interview with Herman Davis, but turned up nothing that I did not already know. Agents had also interviewed Dr. Harry Shapiro of the American Museum of Natural History, who confirmed their suspicions that Foley was not necessarily giving a full account of his connection with the fossils. The report did not elaborate beyond that and it was unclear whether or not Shapiro had.

Shapiro also told the agents that from what he could see in the Empire State Building woman's snapshot, the possibility that she had much of the Peking man collection was slim. He raised hope only for the skull pictured in the upper corner of the box. Shapiro also supplied the name of a widow of a marine master sergeant and suggested that she might be the mystery woman. The agents checked out the woman and found that she was seventy-three years old, considerably older than the woman I had met. Agents added that the woman knew next to nothing about the fossils.

The report also included an interview with Claire Taschdjian. She repeated her description of the boxes in which she had packed the fossils in 1941, but had nothing knew to add.

The report then focused on Andrew Sze. The report read:

A friend, who is a doctor in Taiwan, told him the Peking man fossils were taken from China to Taiwan in 1941 and came into possession of a high government official. When this official died, the fossils were given to his adopted daughter, who is married to a Nationalist Chinese Army officer in Taiwan. She allegedly still has the fossils.

Sze said that at my request he had written his friend seeking further information and the man replied with a Christmas card but did not mention the fossils. Sze said he was then certain his friend would not volunteer further information.

The rest of the FBI report summarized interviews with surviving North China marines, especially those suggested by Herman Davis as having had the closest connection to the fossils. They were all cooperative but told the agents even less than they had revealed to me. One, however, did supply agents with a roster of marines who returned home from the war and had since died. The agents felt the list might give them a line on the Empire State Building woman, if in fact she was the widow of one of the North China marines.

I read the report carefully. Though I felt the agents had performed well and had managed to get to some people I had not heard from, I was disappointed that they had not been able to extract substantive information from Foley and Andrew Sze.

A few weeks later I went to Washington to confer with State Department officials. I met with Francis Tenny, the Director of East Asian and Pacific Programs, who had written the letter assuring Seng that the government had no interest in confiscating the fossils. Tenny, a moderately built man with dark hair and black horn-rimmed glasses, was open and friendly. He pulled a chair out from behind his desk and sat down directly across from me.

Tenny was greatly interested in the progress of the search and asked if there had been any new word from Harrison Seng. I had no news. My real interest in talking with Tenny was to discuss Andrew Sze. Tenny knew nothing of Sze or his part in the mystery, but after listening to the story he promised that the State Department would look into Sze's connections. I mentioned the fact that the last name of the young Philippine woman I had met in China was also Sze and that I had been told that the manager of the Grand Hotel in Taiwan was a man named Eddie Sze. I was eager to find out if Sze was just a common family name or a remarkably active family.

Tenny told me what he knew about the evacuation of

Chiang Kai-shek and the Nationalist government to Taiwan in 1949. He thought that about 2,500 crates of valuables had been taken from the mainland by the Nationalists. It was possible that some of these crates were as yet unpacked and could contain the fossils, but Tenny doubted it. He suggested that I contact Professor James Cahill of the University of California at Berkeley who had photographed and catalogued all of the Nationalist Chinese treasures.

I then asked Tenny the question that had been nagging me almost from the beginning of the search.

"Why should all these people, the Empire State Building woman, Andrew Sze, even Dr. Foley, keep telling me that I'm naive about this whole thing, that I don't realize the danger involved?"

Tenny smiled slightly then leaned back and assumed a serious, businesslike expression.

"It is true, as they say, that you are not the only person involved in this. Certainly the government in Taiwan would like to have the fossils, as would the People's Republic. And then who knows, you offered a reward for them and that can attract a lot of people. When you're dealing with any amount of money there is always some danger involved.

"But most important is the Chinese involvement. You must remember that the Chinese never do anything without a reason. And it is quite possible that both Chinese factions are actively pursuing this. If they are—and of course you're pursuing the fossils—then they are pursuing you."

The rationale stuck with me. I had heard it, though couched in slightly different terms, from both Andrew Sze and Harrison Seng.

Upon my return from Washington I wrote Professor

Cahill at Berkeley and soon after received a reply. Cahill wrote that the crates Tenny spoke of were at the National Palace Museum in Taiwan but, to the best of his knowledge, they contained only objects taken from the Peking Palace Museum—works of art, rare books, and various historical documents. They had been stored in Shanghai, Nanking, and Chungking, among other places, during the war.

Cahill thought all the crates brought over from the mainland had been unpacked, but he was not absolutely certain of this. He said that catalogues of the treasures existed, but they did not cover everything contained in the crates. He believed any archeological material brought from China would have gone to the Academia Sinica, the China Academy, located near the capital city of Taipei.

They [the Nationalists] brought with them, in addition to their library and other research materials, large quantities of archeologically excavated material from the various excavations that they had conducted on the mainland. I suppose that if any human fossils were brought to Taiwan, they should be in the Academia Sinica collection.

But, Cahill concluded, he had never seen them.

Though agents kept in touch with me periodically, the FBI did not play a major role in the search during the weeks that followed. Agents were scouring the country tracking down leads, but their success, if indeed they met with any, was not significant.

While in New York on business that August, I received a message from an agent who asked for a meeting. We agreed upon a time and met at the Drake Hotel. The agent introduced himself and a colleague, then he asked that I refer to him thereafter as "John Zanzibar." I thought the

Zanzibar alias was a bit melodramatic, but I overlooked the point.

Zanzibar started off.

"You probably already know how the bureau got involved with this case but let me review some of the more important elements. Of course, we have a directive from Washington on this, but our only real area of jurisdiction involves the possibility of interstate fraud. One of these people with whom you've dealt may try to extort money from you across state lines."

He went on to cite examples of the type of fraud that could be perpetrated. I was already informed about what he was saying. I was anxious for the agent to reveal the reason for our meeting.

Zanzibar went on. "We've interviewed everybody you've told us about and I would say half again as many people that you haven't mentioned. We've talked to the marines, to Davis, to Dr. Foley, to Andrew Sze, people in Japan, in Taiwan, and we are continuing to investigate leads that have come up.

"But there's one thing I want to make clear, and I want you to remember this: I don't ever want to be cited as the source of any information. This is essential. If I quote somebody in our conversations, that is strictly confidential. The conversations I had with that person were confidential."

The line was a familiar one; I quickly interrupted.

"Now, John, you're doing this for me in a sense, for my search. I'm very much a part of all of this and it seems appropriate that you should tell me everything, regardless of whether it is confidential or not."

"Look," the agent said. "I'm just stating my legal position

here. Whatever I tell you . . ." He stopped and smiled. Then he nodded and went on. "Okay, let's let the chips fall where they may."

He paused for a moment before going on.

"Let's face it, Mr. Janus, I think Washington and New York have asked me to see you because you feel we have not done a good enough job."

"No, no. That isn't the case at all. I simply feel that you haven't gotten all the information I think these people have. This is not a criticism, but look at it this way: I have investigated so many people and heard so many stories and even traveled all over the world looking for clues and yet it comes down to people right here in this country who claim they know who has the fossils or at least know where the fossils are.

"I want to get to the core of the investigation. I think pressure should be put on those who know the identities of these unknown individuals. Don't you think that makes sense? Don't you think we should go to the source of our clues as quickly as possible and stop playing around the edges of this thing?"

Zanzibar listened without taking his eyes from me.

"I understand you," he said. "But look, we have done everything we can. We have had a number of meetings with the key personalities in this case and we've done practically everything but insult them. If they filed a complaint they would probably win it. Don't think we haven't been thorough.

"I even went so far as to say to them, 'Look, are you withholding information for personal reasons, is it because you want to go to China and find the fossils and be the national hero?' "

"But aren't there ways of getting these names legally?"

"No. You must remember what our limitations are. We cannot harass these people, we have no evidence of illegal activity. Some of the information people possess was given them in confidence, and we have to respect their right to withhold it."

"I understand, John, but if you could convince these people that no harm will come to them or their friends in China or in the U.S. . . ."

It was obvious that we had come to an impasse. Zanzibar's companion, who up to this point had been silent, broke in.

"Let me say something here. Mr. Janus, you realize that you're a prominent individual. You're known in this country; you're also known in China. If you went there and talked to a couple of Chinese, whether they were big-wigs or whether they were peasants, you have to admit that the finger would be put on them right away. Maybe Foley is right, you would be involving these people, maybe endangering them if you did this."

I had to agree with him. I did not have to be told about the conspicuousness of an American in China. "My plan, though," I answered, "is to use Chinese-Americans as interrogators. They would be traveling on the pretext of writing a book. Then, with the utmost discretion, they could talk to the people who may be involved."

Both agents saw the possibilities in this idea, even though it was designed to get more information from people the bureau had already talked to. They insisted that agents had pushed their questioning to the legal limits. I felt frustrated, and I wondered whether the FBI was the right agency for the search. It was a law-enforcement group, and it was as yet unclear that any law had been broken. I was left with my original belief that someone, perhaps several people, was withholding information vital to the search. If the FBI could not pry it loose, I would have to.

When we parted, the agents assured me that they would stay with the case and follow up anything that developed. I vaguely acknowledged their assurance, but I knew I would have to solve the riddle on my own, face to face with those who still had a lot more to tell about Peking man.

CHAPTER 11

Two Enigmatic Personalities

No single figure in the entire search remained a greater puzzle than Andrew Sze. The motives of the guarded Chinese-American seemed to be above suspicion. Sze wanted no money, no reward, and most of all, no publicity for any role he might play in the recovery of the fossils. But Sze's initial promise had faded into a dogged reluctance to become involved. And ultimately, he refused to divulge any solid information.

When I felt that I could go no further with Sze, I invited the State Department and the FBI to have a go at it. There was a quality of patriotism evident in Sze's character, and it seemed reasonable that government officials could appeal to him on that basis. He was a relative newcomer to the country, doing well in American business, and he was all too aware that life had not always been that comfortable for him.

FBI interviews with Sze, however tactful or direct, produced nothing. Despite his own request to speak to an official party, he would go no further with FBI agents than he had with me.

In August 1973, I decided to try another approach. I asked a friend in New York to visit Sze at his place of business and see how the man would respond to a total stranger. A suitable alias was arranged as well as a pretense for the visit, and my friend headed for Wall Street.

My friend had little trouble getting past the receptionist at Sze's stockbrokerage firm. Moments later, Sze greeted him in the outer lobby. Sze was smiling and cordial even though he had been given a totally unknown name, and he treated my friend as if he were a long-time acquaintance. He told Sze he was a researcher in Oriental anthropology for a museum in Michigan and was particularly interested in Peking man. Sze seemed not at all surprised and ushered the stranger to his desk in the busy trading room.

Sze was dressed in a dark blue business suit, a conservative tie, and white shirt. He appeared the stereotype of the Chinese-American businessman: the ready smile, the old-fashioned close haircut with no sideburns, the tendency to nod and bow slightly as he responded. Although Sze received only vague answers to how his visitor had gotten his name and exactly what kind of research he was undertaking, he remained affable, and did not hesitate to tell his story.

With a marked Chinese accent, he began to speak in muted tones, always conscious of those around him even though they appeared too busy to pay him any heed. He leaned toward his visitor and looked directly into his eyes.

"Let me tell you the whole story," he said. "Up to now this has been done in the wrong way. I went to Mr. Janus because I thought he was a sincere man interested in doing something good about recovering the fossils."

His brow furrowed as he paused momentarily. "But let me tell you, he is a madman. He is a crusader. He will never

154

find anything. Common sense tells you that he is not going to find anything because nobody will come out like this. He has it in the newspapers, magazines, all over. We have to go about this in a quiet way after all the clamor has died down."

He was intent on getting across his point: his distrust of me, his total disgust with the way the renewed search was being conducted. With little prodding and only the most cursory of questions, he proceeded to tell his story.

He became involved, he said, after he had bought a new house and was attempting to furnish it with Chinese paintings and had contacted a dealer in Taiwan. When Sze talked to the dealer, the man mentioned that he had overheard a conversation in which someone said that he had possession of Peking man. But when Sze read of my search, he decided to come forward with the information to see if anything could be made of it.

His attempts to find the dealer again, Sze said, were unfruitful.

"You see, you can't expect him to cooperate unless this thing is done quietly," he said. "Someday we will do it right. I or someone responsible will get a committee together and we will make an investigation."

But he would go no further than that. He was resolute though amiable, still speaking *sotto voce* and leaning toward his visitor. The only thing that altered his mannerisms was mention of me.

"Use common sense," he insisted. "In thirty years why haven't the Chinese found anything? Why? If they haven't, Janus will not. He will never find anything."

With that, Andrew Sze smiled courteously and asked my

friend to write down his name and the address of his institution. He had told his story and made his point. He smiled again and said he enjoyed speaking to my friend.

The meeting and Sze's story came as no surprise, except for the fact that my friend had so easily pulled off the interview. Andrew Sze was alive and accessible, as close as the nearest telephone. Yet for reasons known only to Sze himself, he no longer wished to be a party to the search or, at least, to *my* search. His protest against the waves of publicity appeared to me to be only a part of his unwillingness to cooperate. I suspected there were other reasons.

Sze had always been the reluctant guest in the search party. He continually spoke about the danger of the hunt, so pointedly, in fact, that at times it seemed as if he had personal knowledge of some past violence. It was that fact, I surmised, or else Sze was a timid man afraid of reprisals from one or the other Chinese government.

One other real possibility in the case of Andrew Sze concerned details in his original story. Not only did he seem to know more than he was telling, he had altered the facts on separate tellings, reciting a version to my friend that was different from those he had related to me and to the FBI. This newest version of an overheard conversation mentioning Peking man rendered his contribution somewhat tenuous. If indeed this version was the truth, Sze's information was skimpy at best and his reluctance to cooperate could have been a way to protect himself.

Still the knowledge Sze seemed to possess about the Chinese and their preferred way of dealing with the lost fossils gave me the impression that he knew something, or someone, and that he was sure of his information. I never lost sight of that possibility, and kept Andrew Sze squarely in the picture. Perhaps Andrew Sze had come forward in the first place because he really knew what had happened to Peking man. Yet now it seemed as if nothing could pull the secret from him.

If Andrew Sze was a source of frustration, Dr. Foley added a generous helping of mystery. Foley was one of the central figures in the Peking man story, an individual who had played a vital role in the fate of the fossils in 1941 and who was available to cooperate in a renewed search thirty years later. With his friend and assistant Herman Davis, Foley was able to fill in details of the last known moments of the fossils and in so doing, to shatter a number of myths surrounding their disappearance.

Perhaps the most striking thing about Foley and Davis thirty years after their ordeal in China was the fact that their relationship had not changed. They were the best of friends and for the last several years, since ill health forced Davis to give up his own business, trusted associates. Yet it was this close, professional relationship that raised questions in my mind. Davis was an outgoing, affable man, quite willing to talk about the North China marines. He had a collection of personal photographs of the Camp Holcomb barracks and of fellow marines, and he actively participated in the division's yearly reunions. There was little that Davis would not talk about in regard to North China; it was probably the most exciting time of his life.

Davis' candor was in direct contrast to Foley's reticence. The doctor was often inaccessible because of his duties and he was difficult to talk to once he was tracked down. I was often reminded of a snapshot I had seen. It was taken in 1945 aboard a troop ship returning from Japan and it showed Foley as a thin, gaunt-faced, deadly serious young man. He had, of course, barely just survived inhuman treatment in Japanese prison camps, but that same expression crept over his face when he talked with me. I felt that he was evading my questions. He continually referred me to an article he had written for the Cornell University *Alumni Quarterly* in which he told the story of his prewar relationship to the fossils and the events of Pearl Harbor Day. What details the story did not supply were quickly recalled by Foley, but only to a point; then he claimed he had no way of knowing anything further.

Foley was one of the few people who encouraged me to make a *cause célèbre* out of the hunt. He thought a publicity campaign and a reward offer were proper and effective means of bringing out the missing elements of the mystery. But his enthusiasm seemed to wane when the mystery began to point toward him.

Even his *Quarterly* article raised some questions that remain unanswered:

> The boxes bearing my name were delivered apparently intact, while the boxes belonging to the marines from Camp Holcomb had been opened and rifled, and their contents delivered in mixed-up bundles. I opened my personal boxes and found that several modern skulls I kept as anatomical specimens, as well as a Chinese Buddha figure, were missing. . . . The footlockers assigned me from Peking, *I had not examined.* [Italics added]

Why Dr. Foley did not inspect the Peking man lockers, why he did not even peek at what was inside, was most perplexing. A number of days passed from the time the Japanese took the marines prisoner to the time Foley himself was imprisoned. During that time he was able to move about, and to prepare for what he knew might be a long period of internment.

He went on to describe his preparations.

Faced as I was with the prospect of internment, which in fact came shortly after and lasted for four years, I decided to distribute the luggage bearing my name in various depositories for safekeeping. Some went to the Swiss Warehouse and the Pasteur Institute in Tientsin and some were placed with Chinese friends on whom I could rely.

Again, unnamed Chinese individuals came into the picture. Foley would not release their names. He seemed afraid, just like Andrew Sze, of endangering the Chinese in their own country. He zealously guarded that information.

Yet Foley showed interest in accompanying me to China on any further expeditions in search of the fossils. His willingness to cooperate in this regard gave me the impression that he did not alone hold the key to the fossils.

Upon my invitation to accompany me to China he said he would be happy to go along if certain requirements were met. He carefully outlined them in a letter headed:

Conditions under which I will return to China and Assist in the Search for the Recovery of Relics of *Sinanthropus pekinensis*.

1. My personal safety must be guaranteed by the Chinese leaders, Mao Tse-tung and Chou En-lai.
2. The search must also include a visit to the graves of our fallen comrades, victims of Japanese torture and maltreatment, buried in graveyards adjacent to the Japanese prisoner of war camps in Woo Sung and Kiangwan (outskirts of Shanghai) in order that I may pay my respects to these heroes.
3. The search will necessarily involve the interrogation of Chinese nationals who assisted me in resisting the Japanese military invaders. I must have the guarantee of the Chinese

leaders mentioned above that these persons will be treated with the utmost respect.

They assisted me in underground activities against the Japanese invaders and are in truth heroes.

If we succeed in finding the relics, no reprisals may be carried out against such persons.

(signed) William T. Foley, M.D.
(In 1941, Medical
Officer U.S. Marine
Corps Detachment
Tientsin, China)

Reading his requirements, I realized for the first time that Foley seemed to have a deep emotional commitment, which contained elements of resentment. If Foley thought that the Chinese who had helped him with his Peking man luggage were to be tortured or killed because of it, he may have believed that the fossils were better off lost to mankind than celebrated in the ashes of his Chinese friends. Such feelings, however confused they might be, clearly played a part in Foley's attitude toward the search and the possible reemergence of Peking man.

I spoke to Herman Davis about the emotional background of Foley's experience in China but he was unable to shed new light on Foley's feelings. When asked about Foley's friends in China—friends who might have information about the footlockers—Davis could only recite the names of several of Foley's Chinese patients and talk of the close ties Foley had had with them. He mentioned that some of them were now living in New York, but he appeared to know little more than that.

Asked whether Foley had discussed his Chinese friends with him in connection with Peking man, Davis said, "At times, but he will not go into detail. He will not tell me who they are because he feels that it would be dangerous for a Chinese to be named anywhere."

It was Davis' opinion that some of the fossils may have been taken from Camp Holcomb but that others may have been left there, thrown in piles of debris the Japanese stacked outside the newly taken buildings. "I'm sure the Japanese handled these things and I'm sure, knowing the Japanese soldier, that he would say the bones were those of some Chinese ancestor and who in the hell needs them. They may have put the bones back in the lockers and some of them may have been shipped out. But I think that some might have been thrown onto the ground around there somewhere."

But though Davis was perfectly open about discussing the fate of Peking man, Foley remained reserved. Almost everyone who interviewed the doctor came away with an ambivalent impression: he seemed to be telling the truth but he probably knew much more; he expressed interest in recovering the fossils but he had no real curiosity about their fate. His facade was not penetrated during any of the interviews, not by me or anyone else. He was a man with a thirty-year history of dealings with the Chinese and Japanese, in some of the most brutal circumstances imaginable, and he seemed to have acquired a measure of the fabled inscrutability of the Orient.

Foley was certain of only one thing: that the fossils still existed.

"Oh yes," he insisted. "Nobody throws anything away in China."

CHAPTER 12

Photographic Evidence

One of the few tangible pieces of evidence I had to work with after my encounters with the Empire State Building woman, Andrew Sze, and the host of other figures connected with the mystery, was the woman's snapshot of the open footlocker. The bones pictured could be compared to casts and, especially, photographs made of the original specimens; they could be viewed by people like Claire Taschdjian and Dr. Harry Shapiro, among others, to arrive at some estimate of their authenticity.

When the snapshot was first shown to experts, they responded with guarded enthusiasm. Many of the bones shown in the footlocker were clearly not specimens of Peking man, they said. Some were obviously femurs, or thigh bones, a bone not among the Peking man collection. Others show portions of joints made up of tissue too soft to withstand decomposition. Some of the bones were lighter in color than others and appeared to be sections of skulls with walls far too thin to be of the *Homo erectus* variety. On the whole, the box of bones had few of the characteristics of the Peking man collection, the experts said. They were generally pessimistic, inclined to dismiss the entire collection in the snapshot as a hoax intended to fool me long enough to relieve me of the reward money.

A typical and particularly disheartening opinion came from Dr. Richard Leakey, Director of the National Museum of Kenya and the son of the late Louis S.B. Leakey, the anthropologist who discovered the famed *Zinjanthropus boisei* remains in Tanzania's Olduvai Gorge in 1960. After being shown the picture and given a detailed summary of my findings in the Peking man search, Leakey replied by saying that he thought the entire endeavor was a waste of time.

In a letter dated July 13, 1973, he wrote,

Quite frankly I do not believe that the fossils you are looking for will be found. It is my firm impression from talking to colleagues, particularly in the older generation, that this material was lost during the last stages of war and it seems certain that they are not in Europe, the United States, or the Far East.

Leakey went on to say that the recovery of the fossils would be of the greatest importance, but he discouraged a continuation of the search. He had even worse words for the Empire State Building woman's box of bones.

The unidentified woman appears to have given Mr. Janus a photograph of some modern bones. I am astonished that anyone felt that the photograph represented possible fragments of the Peking collection. Anyone who is familiar with the details of the material that was lost could tell at a glance that the photograph is not of that collection.

Other experts were not as quick as Leakey to close the door on the snapshot. Their interest was aroused mainly by the skull pictured in the upper right-hand corner of the footlocker. This specimen was unlike the others pictured, and was situated in the box in a way that made it even more mysterious. Dr. Harry Shapiro first singled out this object even though he, like Leakey, quickly ruled out the rest of the pictured bones as too modern to be Peking man. The skull, and only the skull, was strikingly like the original specimens.

Shapiro's observations were supported by Dr. Glen Cole of Chicago's Field Museum of Natural History. Cole found it significant that the skull was positioned in such a way that it was impossible to get a complete look at it and better verify its authenticity. What the anthropologists could see of the skull, however, raised important questions.

The possibility that the skull was authentic was enough to keep me on the trail of the woman, and I continued to seek more opinions. Some of these came from anthropologists who visited Chicago in the summer of 1973 to attend the Ninth International Congress of Anthropological and Ethnological Sciences. I was invited to lecture about the Peking man search and to participate in a panel discussion. Among the anthropologists there was G.H.R. von Koenigswald, one of the original researchers who viewed the Peking man collection in China. Von Koenigswald, however, dismissed the drama of the hunt and proclaimed unequivocally that the fossils were lost in China years ago.

Many experts openly disagreed with von Koenigswald and were fascinated by the search. One of these was Phillip V. Tobias, a South African authority on *Homo erectus* and Peking man. Tobias studied an enlarged copy of the snapshot after hearing my story. It did not take Tobias long to find what he was looking for.

For him, as with Leakey and the others, most of the bones pictured were easily disqualified as belonging to *Homo erectus*, especially those with the wrong (or at least different) coloration. The skull drew Tobias' attention. With a magnifying glass he identified what he thought were signs of damage: fracture lines and a small area of missing bones near the *bregma*, the point at the top of the cranium where the two parietal bones of the skull meet the frontal bone. He also said the skull looked as if it had been flattened or crushed, a feature, he said, common to many fossil fragments.

But there were other features, judging from the picture alone, which marked this skull as possibly that of *Homo erectus*. At the back of the skull, the point pictured lowest in the box, Tobias pointed out evidence of a ridge of bone called the occipital *torus*, a structure common to *Homo erectus*.

There was also a suggestion on the top of the pictured specimen of a supraorbital *torus*, a bony ridge above the eye sockets. But the picture, he stressed, was least effective in pointing this out.

A third feature of *Homo erectus* evidenced in the photo was what Tobias called the sagittal reinforcement system of the skull. That feature was described by Franz Weidenreich in his definitive study.

Tobias also discerned that the skull appeared to have a marked fullness in the mastoid region (just behind the ears). This was a striking feature of the squat skull of early hominids. Such a skull structure demonstrated that the upper part of the brain was not fully expanded as it was to become in later man and finally in *Homo sapiens*. Tobias was quick to add, however, that the features he noted in this skull were peculiar to most skulls of *Homo erectus*, not only to Peking man.

Tobias emphasized that he was describing only what he could make out from a less-than-high-quality photograph. Nonetheless he was taken with the search and the challenges it presented. He said that there was much work to be done on the original specimens should they be recovered. Tissue structure, chemical makeup, age, bone superstructures and force lines, among other concerns, all remained to be analyzed.

"We have new techniques for analyzing the anatomy," he said. "We can find out a lot more about the functional meaning of the skull. Weidenreich's work was excellent, but it was by no means the last word. It was only the last word on a descriptive basis."

Tobias' interest and evaluations inspired other experts to take another look at the photograph. One of these was Dr. William W. Howells of Harvard University. At an earlier viewing, he had cast a negative opinion on the woman's collection, but at Tobias' request he took a longer look. His reaction was now altogether different.

In a letter to me on September 21, 1973, he explained his new opinion.

Dear Mr. Janus,

I have something surprising to say, and a good deal of crow to eat. I had always thought the box of bones was so much rubbish, because it did not correspond to the Peking remains, and it was difficult to check the skull from the publications of Weidenreich, obviously for others as well as myself. I had also ruled out the possibility that these were the Upper Cave remains, after looking at the published pictures of skulls.

Until Phillip Tobias was looking at the photograph with us, and suggested considering the skull alone, it had not occurred to me to do so. This was a mental block, and very foolish. I have now got out our very good casts of the Locality I skulls, and I find really compelling details present in the photograph and on the cast of skull X I, also known as skull II, Locus L. I enclose some Polaroid pictures of the cast made this morning. The correspondences are even more convincing when the cast itself is viewed.

So I now think that the box contains either the original of this skull or a cast such as we have. I do not see how a cast would be involved. What all this means as to the whole lot of the original material I cannot imagine.

This is, of course, a big surprise to me, although it is my own fault. I was seeing things in the picture I did not notice before.

> Best wishes,
> W.W. Howells
> Professor of Anthropology

In less than a month Tobias became even more convinced of the authenticity of the skull specimen. In a letter to me on October 17 from South Africa, Tobias described results of further study of the casts he had in his possession. He was most concerned with the thickness of the cranial bone, a pivotal element in the skull of *Homo erectus*. This is a characteristic that would normally be difficult to ascertain from a photograph, yet he had detected the great thickness of the skull through an unfilled gap in the cranial vault—as photographed obliquely from above and behind.

"Now the cast of that skull which I possess," Tobias wrote,

and which I believe is the only cast of the skull available, is the restoration by Weidenreich of Skull II, a female from Locus L, *and the gaps between the cranial bones have been filled up with plaster*. This means that, because the fracture gaps are filled in, one cannot by looking down on the upper surface of the cast of this skull see the thickness of the cranial bones.

He went on to explain:

Of course, there is a possibility that another cast was made, in which the fracture gaps were not filled in, so that by looking obliquely downwards, from above and behind, one could have seen, even in such a cast, the thickness of the cranial bones. However, I do not know that such a cast was ever made for this particular cranium; the only one I have ever seen is the kind which I possess in my departmental collection—i.e. with the gaps filled in.

This therefore makes it highly likely, in my opinion, that the skull in the footlocker is an original Peking man skull, namely the one identified by Bill Howells of Harvard, and not a cast of it.

For Tobias, this bit of circumstantial evidence swung him even closer to the opinion that Peking man was still in existence. He wished me the best of luck.

PART FOUR

Back to Asia

CHAPTER **13**
The Search in Taiwan

From the moment Andrew Sze mentioned his unnamed friend in Taiwan, I knew I had to visit the Nationalist Chinese. Sze had not been the only person to suggest that someone in Taiwan possessed potentially useful information. That the fossils somehow had fallen into the hands of Chinese who fled with Chiang Kai-shek in 1949 was too strong a possibility to remain uninvestigated. If private individuals did not have the fossils, or possess crucial information about them, perhaps officials in Chiang's government knew something. The Nationalist Chinese had been accused of taking everything that was not nailed down when they evacuated China. More importantly, the boastful (though perhaps uninformed) letter of Y.T. Chen of the Nationalist Chinese Embassy in Washington stated outright that the fossils were moved by Chiang's government to the island.

My decision to make the trip was finally taken in September 1973, when I received a letter from Gerald Beeman, the ex-North China marine and Cleveland Museum of Natural History curator who had been cooperating with the search. In a letter dated August 30, Beeman wrote, "If information I have just received is correct, you will find the bones of the Peking man in the Nationalist Chinese Museum in Taiwan along with the other Chiang treasures which the Nationalists took when they evacuated the mainland."

Beeman went on to describe the way he had received his new information.

At a cocktail party given by a Japanese officer who was about to return to Japan, I came about this information in a rather strange way. This officer is the same fellow through whom I tried to obtain information on the Japanese soldiers who participated in the capture of Camp Holcomb in Chinwangtao.

Among the guests at the cocktail party was a Chinese Nationalist officer, who is also serving as a liaison officer at Wright Patterson Air Force Base. His name is Colonel Weh and he has been in the Nationalist Air Force about 30 years. He has served three tours of duty in this country, including his initial training during the early part of World War II.

I naturally struck up a conversation with him and in the course of the conversation I indicated that I was involved in the search for the Peking man. I asked if he knew what I meant by Peking man. He stated that he did, that he had read that a Chinese Nationalist general had brought a box of bones which he had not opened with him from North China and that they were probably stored along with other items from the mainland in the National Museum.

To Beeman, the officer's story seemed quite plausible in the context of the outbreak of the war and the situation in China in 1941.

The Japanese must have recognized that the only means of disposal of items looted at Chinwangtao was through the Chinese. Also, the only Chinese in a position to buy them, of course, were the wealthier Chinese, who were for the most part Nationalist in political philosophy. Furthermore, just as the bones originally were given to the military . . . the marines . . . for removal from

China in 1941, so they might also logically have been given to the military . . . the Nationalist army . . . for removal from the mainland in 1949.

You will recall that the Chinese Nationalists have been very reluctant to display the treasures removed from the mainland, probably out of fear that they might be stolen since they are well aware of the fact that the Chinese Communists would like to have the treasures returned.

This protectionist attitude on the part of the Nationalists could very well account for the mystery of the Peking man.

The story and subsequent conversations with Beeman were enough to make me book a flight for Taiwan. I wanted to check this new lead myself. In further talks with Colonel Weh, Beeman said the officer spoke with greater caution, giving Beeman the impression that he knew more than he had told. It appeared to me to be another case of an informant retreating from his original position once he realized that his story would be checked out. In any case, the colonel's story was yet another finger pointing in the direction of Taiwan.

Before I left, I contacted various people involved in the search and asked them if they had any suggestions about whom I should see in Taiwan and what investigative approach I should take. One of these was Andrew Sze. I told him I planned to sit in the Grand Hotel in Taipei and wait for his friend or contact to communicate with me. All he had to do was relay my whereabouts. But Sze insisted he was unable to play a role in the search. "If I could supply you with these names," Sze said, "then I myself could be a hero."

Disappointed with Sze's continued refusal to cooperate, I looked to Professor James Cahill of Berkeley. He gave me some suggestions as to whom to talk to and where to go in Taiwan, but again he insisted that he had no useful information about the whereabouts of the fossils.

I also called the State Department and informed them about my trip. I wanted them to know because I felt they could help me with the purely technical aspects of getting around on the island. I also thought that State Department officials in Taiwan would be insulted if I came to the island without their knowledge and they later found out about it from other sources. These were formalities I felt I had to go through if I were to make the most of the trip.

I looked forward to seeing Taiwan again and staying at the impressive Grand Hotel. After I was settled, I notified the American Legation and arranged to meet with embassy officials. I first talked to Harvey J. Feldman, a pleasant, attractive young man who was the embassy's counsel for political affairs. Feldman had been briefed about me and the search by State Department officials in Washington. He was most cordial and said he wanted to be as helpful as possible but I should understand certain political facts about the United States–Nationalist Chinese situation.

Feldman said that I had to be very careful in my conversations with the Nationalists because they believed I was an agent of the People's Republic. They knew that the Communists had originally asked me to look for the fossils, and to them that explained my zeal for the project. Feldman also said the United States Embassy would help me in every way possible, but that it did not want the search to look like an official effort. Feldman emphasized that officially the United States still recognized Taipei and not Peking, even though diplomatic realities might suggest the opposite.

I told him that in view of the circumstances he described, I would change the conditions of the reward. I was simply interested in recovering and authenticating the fossils and was dropping my pledge to return them directly to the Peking Man Museum in Choukoutien. I would leave the final deposit of the specimens up to the governments involved. I had not begun the search with political motivations and I had no interest in cultivating them now.

Feldman approved of this sentiment. He added that he felt the Nationalist Chinese would not have kept it a secret if they had taken the fossils in the first place. They had shown the treasures they removed from China in museums built expressly for that display. They would not have kept Peking man hidden. He also said that if the fossils did turn up in Taiwan, the Nationalists would under no circumstances return them to Peking.

I also spoke briefly with William H. Glysteen, the embassy's chargé d'affaires, and was then directed to Ambassador George Yeh of the National Republic of China in Washington. Yeh was considered the elder statesman of the island and deemed the most vital contact for me to establish in Taiwan.

In impeccable English, Yeh greeted me most graciously in the government building the following morning. Though he appeared to be in good health and extremely alert, I learned that he had recently suffered a heart attack that had forced him to transfer his offices from the sixth to the first floor. His greeting, I thought afterward, was quite similar to those I had experienced in Peking. Yeh was very interested in Peking man, and he told me that he believed the fossils had been lost at sea. In an animated manner, gesturing with his hands as he talked, he went on to recite the familiar story of the U.S.S. *President Harrison* and her flight from the Japanese in 1941. I discussed the story with him and explained why I thought it was not true.

When he asked me how he could be of the greatest help in the search, I revealed my suspicions that the fossils might be in Taiwan, perhaps in the National Palace Museum collection itself. Yeh was surprised but agreed that almost anything was possible. I liked the old man's attitude, he seemed to acknowledge the possibility that the most unlikely events, the most improbable factors, prevailed in such a complicated search. Yeh then picked up a telephone and arranged for me to meet the director of the Palace Museum, Dr. Chiang Fu-tsung, and the director of the National History Museum, Dr. Ho Hao-tiem.

The ambassador also suggested that I hold a press conference or, if I preferred, a series of interviews with the local press. I agreed, although I was unsure exactly what purpose a press conference would serve at this time. In ten minutes a representative of the government's press office arrived and promised to arrange a conference. Yeh and I chatted for some time, both of us describing various projects we were involved in besides the Peking man search. I did not learn anything new from him, but the statesman had assured me that some of my leads might be realistic possibilities.

The target of the Taiwan hunt, according to my sources, was the awesome National Palace Museum. It was here that Cahill said all the crates taken from China in 1949 were stored. It was here also that Colonel Weh had said the fossils could be found. I approached the museum and its director with anticipation.

The Palace Museum was an overwhelming Chinese

structure. It reminded me of Leningrad's Hermitage Museum. But I was not to see the museum immediately; first I was escorted to Dr. Chiang's spacious and well-furnished office. With a group of museum staff members standing by, Dr. Chiang toasted me for my efforts and thanked me on behalf of the Palace Museum. His effusive praise, however, made me feel somewhat ill at ease since I had come to the museum expressly to confront Dr. Chiang with the possibility that the fossils might be inside this very building.

I tried to get to this point tactfully.

"You know, Dr. Chiang," I began, "in this search I've come across the most bizarre evidence. It has led me to suspect everything, to follow up even that which looks most unlikely to be true. I simply have to investigate everything."

Chiang nodded and smiled, and the others present seemed to be in full agreement.

"Then I hope you won't be offended if I tell you that it has been suggested that the fossils may be here in the Palace Museum."

Dr. Chiang threw his head back and laughed; the others followed suit. I was grateful for their sense of humor.

"Where do you think they are?" Dr. Chiang replied.

"Perhaps among the four thousand cases?"

Dr. Chiang nodded and said he understood why I might draw this conclusion.

"Let me show you the cases. You can see for yourself," he offered.

The director led me and the others through the museum toward the rear of the building. I could see that the massive building had been built at the base of a mountain, butting up against it in such a way that there was no rear entrance. At the building's rear stood a pair of large steel doors which led directly into a causeway. The entourage walked through it into a well-lit concrete tunnel. I wondered to myself if any other American had been inside it before. I was looking into the middle of a mountain, where centuries of Chinese treasures had been stored.

"Here they are," Dr. Chiang said. "All the crates—more than four thousand of them—that were brought to Taiwan."

We walked slowly through the tunnel, a space about eight feet wide and fifteen feet high. Wooden crates were stacked on all sides, each marked and apparently catalogued. The crates left a walkway large enough for only two people to stand side by side.

As I walked, slightly bumping the shoulder of Dr. Chiang, the museum director slapped his fists against the passing crates.

"Empty! Empty!" he proclaimed, striking each of the hollow boxes in turn.

We walked further and I realized the tunnel continued into a maze of passageways, all stacked with crates, stretching as far as I could see.

"Would you like to look in them yourself?" Dr. Chiang asked.

I nodded but felt somewhat sheepish. I lifted the lids of some of the crates stacked nearest to me and found that they were indeed empty. I obviously could not look into every one of them, but I had satisfied my own curiosity and permitted Dr. Chiang to make his point. I had no reason to believe that Dr. Chiang was lying to me.

"We've opened and catalogued everything. The fossils are not here and never have been," Dr. Chiang said. "If they were, we would most definitely show them."

We walked on for some time past the unending stacks of

crates and cases. The lighted tunnels appeared to go in all directions. I asked if I could photograph the tunnel but Dr. Chiang quickly refused.

A few moments later, I said, "Okay, I see the evidence. But because so many people think they're here, would you look again? It may sound presumptuous of me to say it, but maybe the fossils are somewhere here and even you do not realize it."

Dr. Chiang smiled, then nodded in agreement. "Yes, yes. We'll look through them again. We'll do that just to satisfy you."

We left it at that, and shortly after, the heavy steel doors closed behind me and I was back in the museum proper.

Dr. Chiang insisted that I see the museum and its treasures. I spent the rest of the day on a guided tour and saw some of the most exquisite Oriental art ever created. It was, the guide pointed out, the heritage of the Chinese people.

I next went to the China Academy, the Academia Sinica, the Taiwan version of what was once China's main center of anthropological research. Professor Cahill wrote that the Nationalists had brought their library and research materials here, along with large quantities of archeological specimens excavated in China. The academy would have been the appropriate spot for Peking man. Furthermore, the academy's senior scientist was Dr. Li Chi, the seventy-seven-year-old anthropologist who had been looking for Peking man since its disappearance.

I was immediately impressed by Dr. Li, a frail man of about five-foot-eight. Though he was clear-eyed, his hearing was poor and he seemed to shuffle as he walked. He was highly regarded in his field, however, and had been a friend of Davidson Black at Choukoutien in 1927. Though not personally involved in the excavation, he had been closely connected with the early research. The old scientist, who earned his anthropology degree at Harvard in 1923 and spoke perfect English, told me that he had been both scientifically and emotionally involved with the fossils almost since the first tooth was discovered, nearly fifty years before.

He spoke with me in the presence of the academy's director.

"I think you're searching in the wrong place. I was in charge of trying to locate the fossils after the war and ever since, and I think they are in Japan."

I listened closely as in a vigorous but soft voice he went on to describe the investigations he had undertaken in Tokyo after the war. He had spent more than a month looking through artifacts and fossil remains which the Japanese armies had taken from China, but he had discovered nothing.

"I feel sure the fossils are in Tokyo. You should go there."

"Perhaps I will," I replied. "But on another mission. Right now I want to pursue the search right here."

The old man smiled. "Essentially, Mr. Janus, you are looking for two men. One is maybe a million years old and the other one has him. Is that right?"

I smiled and the two Chinese laughed.

"But you must realize that there is no possibility that the fossils are with us," Dr. Li said.

"I think I agree," I said. "But I am looking into the impossible. And because of it, I must ask if it is possible for you to again search in the academy. I think it would be important."

Dr. Li looked over to his colleague. They smiled.

"This is a very persistent young man," the academy director said.

I was amused by his perspective on my age.

"The very fact that it sounds preposterous for the fossils to be here makes it all the more possible in my mind."

The two men nodded and seemed pleased with the statement which apparently struck a responsive chord. They agreed to look again through the academy's holdings.

The three of us chatted on for about an hour. Dr. Li expressed some suspicion that the fossils had been lost at sea, and I explained why I believed that story to be an impossibility. He appeared surprised at first but seemed convinced by my information about the U.S.S. *President Harrison.*

After elaborate expressions of thanks for my efforts, the two men said goodbye to me and I drove back to Taipei.

My investigation of Taiwan had turned up two dry wells so far. I had confronted prime sources and drawn blanks, managing once again only to negate persistent rumors. It was a necessary task but hardly a gratifying one. I had come to the other side of the world to find the fossils, not to lay rumors to rest.

Yet my last stop, at the National History Museum, would change all that. I had no way of knowing that this expedition would suddenly shift the direction of the search and lead me to another side of the Taiwan mystery.

As I walked into Dr. Ho Hao-tiem's office I was confronted by a large painting of Peking man, the very statue I had posed next to in the Peking Man Museum in Choukoutien. In an outer office, I spotted a model of Peking man, again similar to the Choukoutien exhibits. I was told that the two pieces were part of an exhibit on Peking man the museum was planning. I wondered whether such an exhibit had been expressly planned for me or if it was simply a coincidence.

Dr. Ho was eager to see me. He thanked me for my search, speaking as though for the entire scientific world. I accepted his effusive praise but was struck by his formality. A short, stout man in his fifties, he was dressed in a dark suit with a stiff, uncompromising collar that fitted his manner perfectly. Only his kudos were unrestrained.

"Of course, if you find them, then you will certainly bring them here!" he said. It was clearly an attempt to be facetious, but there was a distinct element of sincerity in his statement. I did not respond to it, remembering what Feldman had said about the sensitive politics involved.

The two of us spoke through an interpreter, and discussed the fossils and the theories surrounding their disappearance.

"I do not believe they are in Taipei or anywhere in Taiwan," Ho said. "But the story about a general taking the fossils is true—I know that. I know the details about this man and I know his name."

At this, I sat up and listened intently, thinking that possibly my theories and the information from Gerald Beeman were correct.

"Where is he now?" I asked.

"Some say he is in Hong Kong. I've done some investigating on this thing myself and I know that he is not here," Ho replied.

"What is this general's name?"

Ho paused and smiled. He began again slowly, shaking his head.

"My friend, forgive me, but this is a delicate political problem. There are a lot of people involved, some of them people close to me. I simply cannot give you the name. But if you go to Hong Kong, I will arrange to have him or his representative contact you."

My mission in Taiwan, it seemed, had not been fruitless but once again the trail had forked off in a new direction.

Ho went on to talk about other leads and mentioned that the museum was very interested in the reward. He then suggested a press conference as had George Yeh, and insisted that I hold it in the museum. This seemed to me also to be the best possible place for it.

The following afternoon I returned to the museum for the conference. I was met in the courtyard which was buzzing with activity. Television cars were parked outside and a large number of people were milling around. Once inside, I was confronted by a gathering of about fifty people in the conference room, most with pencils and paper in hand.

Dr. Ho supplied a brief history of the fossils and their importance, and some background about how I had become involved. I then gave a brief talk, lasting about fifteen minutes. The thrust of the questioning that followed centered on what I would do with the fossils if I found them in some place other than Taiwan.

I earnestly tried to explain my plans. "I want to stress that I am not an agent of the Peking government, nor of the U.S. government, nor of any investigative agency such as the CIA or anything else. I'm here as a private individual and I have no political connections. I'm only interested in locating the fossils."

At first the reporters did not seem convinced; they apparently had been predisposed to believe that I was an agent of some kind. But my insistence on my independence gradually seemed to make an impression. Some reporter observed that my present position differed from the stories that had come out of Peking of my pledge to return the fossils to Choukoutien. I agreed with the charge but replied that my real and only goal was to recover the fossils and authenticate them. I would leave it up to government officials to decide where they should go. The reward for the recovery of Peking man still stood, but I raised the offer to $150,000.

That night I was the guest of honor at a small reception given by Dr. Ho. About a dozen persons attended the dinner, which was very similar in style and atmosphere to the reception given our group more than a year earlier at the Peking Duck Restaurant. We sat around a table and toasted one another with Mao Tai. Dr. Ho reiterated his admiration for me and said again that the entire scientific world was indebted to me. Other toasts were made and responded to before the lavish meal was served. After we had eaten, Dr. Ho repeated what he knew about the renegade general who had gone to Hong Kong. But he again refused to name him, maintaining he did not want to get in trouble with the Chiang government. Then he wished me luck in Hong Kong and promised he would do everything possible to get the general to contact me. I thanked him in the spirit of the toast. I had long ago resigned myself to the frustrations of the search.

There was little I could do but trust the Nationalist Chinese and head for Hong Kong. Before I left, however, I looked up Eddie Sze, the manager of the Grand Hotel. I asked him about his name and inquired if he had a relative named Andrew in the United States. Eddie said he had no kin in America and that Sze was a common Chinese name. I thanked him and thought once again of Andrew Sze. It appeared that Andrew's tracks were covered; his secrets still belonged to him alone.

The following day, October 5, I paid my respects to the various officials on the island, including the embassy personnel and Ambassador Yeh, and boarded a plane for Hong Kong. The trail was growing fainter; I did not even have a name to go on.

CHAPTER **14**

Hong Kong and the Golden Triangle

Hong Kong was where it had all started. I remembered my short stay at the Peninsula Hotel, the meetings with Mr. Nine, the trip to Macao, and the crossing into China. Now, more than a year later, I was back, hoping my search might end where it had begun. I could do little but wait for Dr. Ho to come through for me.

Once again the hotel was expecting me and had arranged a suite. But before I could get to it I was confronted by three reporters who had been tipped off about my arrival. I pleaded fatigue and promised to fill them in later. That evening, however, I received a call in my room from a reporter for the South China *Morning News* whom I remembered from my earlier stops in Hong Kong.

I told him of my trip to Taiwan and what Dr. Ho had told me about the Nationalist Chinese general. He was familiar with such stories and said that if the general was anywhere in Hong Kong, he was most likely in Rennie's Mill, a small settlement just south of the city. Nationalist Chinese military personnel had settled there after Chiang had been routed from China. The community was completely separate from Hong Kong and had its own government.

I took down this information but decided not to investigate immediately. I would wait instead for Dr. Ho's general to get in touch with me. I was sure my presence was known; Hong Kong papers had reported my arrival and the story of the search on page one as soon as I got off the plane. They were obsessed with the story. I only hoped the general was equally interested.

But after two days of waiting, I had received no word so I decided to go to Rennie's Mill. It was a Sunday morning and I debated about taking along a reporter to document my findings. I ultimately decided against it and, in one of the hotel's cars, I set off with only the driver and an interpreter. The car headed south out of Hong Kong and after about a half-hour's drive ran into winding, dirt roads that stretched into the mountains near the coast. Neither the driver nor the interpreter knew exactly where the village was.

After driving ten minutes through the hills, I spotted a Nationalist Chinese flag waving at the top of a nearby cliff. The three of us got out of the car and climbed the hill but found only a solitary, empty building and the flag. Soon after, however, a group of laborers came up the road and the interpreter asked them about Rennie's Mill. The men said we had not gone far enough, the village was farther down the road along the coast. Another ten minutes of driving finally brought us into a clearing; we were at Rennie's Mill, a settlement of about fifty or sixty buildings. The Nationalist Chinese flag flew everywhere.

As our large car pulled into the center of the village, hordes of children and dogs gathered around it. The reception reminded me of my visits to small villages in Greece where, as a stranger, I immediately became the biggest single attraction in town. I was told that everything was closed in the village but that I should go to the village office building at the bottom of the hill. Since there was no road, we had to walk. We started out with our escort of children and dogs. The morning was hot and I found myself breathing hard at the end of our mile-long trek.

My reception there, however, was hardly promising. The office was closed and I was told I had to go to a smaller office in a nearby house. There I was introduced to three men who said they were village officials. They told me the village leader was not there and would not be available until the next day. I explained to them why I was there and asked if they knew about Peking man. They did, they quickly replied, but upon further conversation it became clear that they knew only what had been reported in the Hong Kong papers about myself during the previous two days. They knew nothing about the mysterious general but suggested I talk to the village leader about it. His name was Kui Sik-kau. I decided I could do nothing but promise to return the next day to talk with him.

I began the tortuous trip back up the hill to the car. It was exhausting; I found myself leaning on the shoulders of my interpreter most of the way. It took nearly an hour to reach the car, and I could barely stand up when I got there. I had seen Rennie's Mill, though I dreaded the thought of returning the next day to trek back down the hill.

That afternoon, however, I was spared the trip—I received a call from Kui Sik-kau, who apologized for not being in the village that morning.

"Yes, I know about the fossils," Kui said. "But they are not in the village. Tell me, is the reward still on? Are you still serious about the money?"

"No question about it. The offer still stands."

"Okay," Kui said, "I promise you I will make a search of the village. But I know who you are looking for. That man did not come here, Mr. Janus. He went to the Golden Triangle."

It was the first time I had ever heard that name. I pressed Kui to elaborate.

Kui's story was sketchy. The Golden Triangle was located in northern Thailand along the borders of Burma and Laos. Several of Chiang Kai-shek's Kuomintang generals had fled there with their regiments in 1949. The small country of Thailand was fearful of attack by China and the Kuomintang generals were able to play on this fear by making a deal with the Thai government. They would settle in the Golden Triangle and act as border guards against the Chinese. The only thing they asked in exchange was control of the region's opium trade, for the Golden Triangle was one of the most fertile opium-growing areas in the world. With this shabby but convenient bargain in hand, the generals took over the Triangle and have lived there ever since.

"That is where you will find the man you seek," Kui said. "Go there and see him and you may find the answer to your quest."

I was taken with the new story. I told Kui I would go on to Bangkok immediately and try to make a connection with the general.

"But be careful," Kui warned. "These men are very dangerous."

I took off for Bangkok the next day. One thing had led to another; I had no idea where my search might end. I had been to Bangkok ten years before, but I remembered very

little about the city other than the name of the hotel wher
I had stayed. I checked into the Arawan, an old, establishe
house which was as magnificent as any hotel I had been ir

As in Taiwan, I first contacted American embass
officials. Because I had no contacts in Thailand and n
specific destination other than the area of the Golde
Triangle, I was more dependent on the embassy here than
had been in Taiwan. I conferred with William Kushlis an
John Barnes, the latter a political adviser to the embassy.
got right to the point: How could I get to the Golde
Triangle?

At the mention of the area, both men became visibl
concerned. Why this area? What had I learned that woul
send me there? I explained my story. The two men listene
to every word, then wasted no time in telling theirs.

"You couldn't have picked a more sensitive area here i
Thailand," Barnes said. "Our mission here for the past fev
years has been concerned with what we are trying to do i
the Triangle."

He went on to tell me a checkered history of dealing
with the drug-running generals in that area. No amount c
cajolery or compromise could put these men out of th
business. The stakes of the opium trade were simply to
high. The United States Embassy had attempted to make
deal with the generals in 1972. They had offered them
million dollars in exchange for a bonfire fueled with th
year's opium harvest. Barnes said the embassy had followe
through with its part of the transaction, and fires had bee
staged. But it turned out that what was going up in smok
was not opium at all, but hemp and other flammabl
materials. The opium traffic from the Triangle continued t
flourish.

190

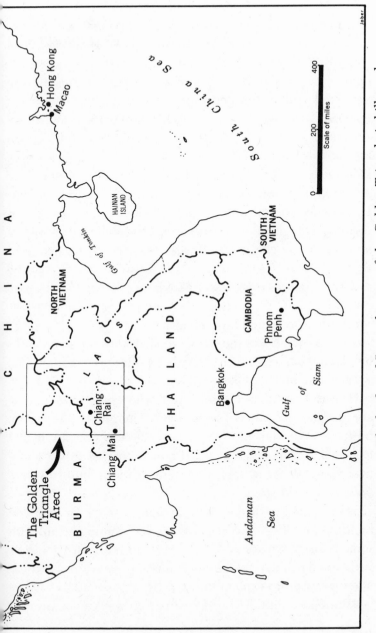

South East Asia, Thailand, and Laos; the area of the Golden Triangle is hilly and almost inaccesible.

Barnes and Kushlis described the Triangle as a dens
hilly region that was almost impenetrable. It was north
Chiang Mai, the nearest population center, and accessib
only by air or horseback. The Kuomintang generals lived
scattered settlements along with their regiments. There w
constant warfare between each group over control
various areas of the opium region. It was doubtful, accor
ing to Barnes, that the Nationalists were still a viab
fighting force. But because of the availability of guns an
ammunition, they were heavily armed, considered e
tremely dangerous, and regarded individually as ferocio
fighters. The generals were extremely mobile and moved
and out of the enclave with ease. Because of the nature
the terrain and its inhabitants, the Triangle was considere
even by Thais as a lawless area where one went at his ow
risk.

Kushlis was more blunt about it.

"The generals are the most crooked, unscrupulous, rutl
less men we have ever dealt with." Under no conditions, h
warned, should I try to find them.

It was then that I discovered who they were talkin
about. The general considered the kingpin of the opiu
business and who, the officials speculated, was the man
was probably looking for, was Li Wen-huan. Another figui
was General Tuan, less important than Li but still ver
powerful in the area.

"If you made a deal with these men, perhaps for your on
hundred and fifty thousand dollars, you would end up wit
some bones," Barnes said. "But I would not vouch for the
authenticity under any circumstances."

He went on to explain the embassy's position.

"We can only tell you the situation, we cannot prever

you from going up there or from making an attempt to communicate with the generals. But if you go up there, they will think you have come in an official capacity because you're an American. We can't have that. We don't want to give the impression that we're cooperating with them. The Thai government, however, has an interest there. They've been working with us in our drug-control program. But they are more interested in some kind of a defense against communism on these borders."

Barnes and Kushlis went on to talk about the political circumstances, the unique situation in the Golden Triangle, its location, and its plentiful opium supply. Danger was a way of life there, they repeated, and the appearance of an American such as myself could be tragically misinterpreted. Barnes then mentioned that he knew I had been active in an American antidrug organization and said that this could severely damage my position with the generals. He was referring to U.S. People for the United Nations, a group I had helped form for the purpose of educating farmers in opium-producing areas in diversifying their crops and ultimately getting them to stop growing opium.

"We know you're identified with the organization and the generals could easily find out," Barnes said. "They could draw any number of conclusions about your motives. That's one more reason why we would prefer you not go up to the Triangle."

I had anticipated this reaction, yet I was not sure how to get around it. I had every intention of either going up to the Triangle myself or making some kind of indirect contact with the generals.

My problem was once again solved by a reporter. I was contacted that night by Steve Van Beck, a journalist with the *Bangkok Post*, who said he was familiar with the situation in the Triangle and was eager to help me make contact there. Van Beck wanted to write a series of stories about my search, from the moment I entered the Peking Man Museum to my arrival in Bangkok. He told me that what the embassy officials had said about the area and the generals was true. As a grisly confirmation, the *Post* that day carried a story about the roadside hijacking and murder of two men near Chiang Mai in the Triangle.

Van Beck said he knew of a freelance journalist who had extensive contacts in the area and might be able to serve as a liaison between me and the generals. The man's name was Hardy Stockmann, and he could only be reached, as far as Van Beck new, at a place called Pat's Tavern in Chiang Mai.

The idea of working through an agent seemed like a good one but I checked it out first with the embassy officials. They had heard of Stockmann and said they would ask their man in Chiang Mai for his evaluation.

Stockmann was known to the Chiang Mai officials as a soldier-of-fortune figure with a working knowledge of the Golden Triangle and significant contacts in the area. The Chiang Mai official had no objections to my using Stockmann as an intermediary.

I assured Barnes that I would not take it further than that. I would attempt to contact Stockmann as soon as possible. The embassy people were relieved.

I then got back to Steve Van Beck.

"I've already talked to Stockmann," Van Beck said. "He says he won't talk to you until he checks you out. I guess he got the word already and he told me he wanted to make sure you're not an agent and that you aren't interested in drugs. He doesn't want you to foul up his reputation there. If you're legitimate and can provide the money, he said he would deal."

Van Beck suggested I write Stockmann a letter and let him make the next move. The trip was beginning to exhaust me; I had been in Asia almost three weeks. I knew that if I left Bangkok, I would have to rely on Van Beck to act as my representative.

"You write the letter. I'll be here and keep in touch with Stockmann," Van Beck said. "If there is really something, if he can make contact with your general, you can come back. Trust me to handle things for now."

His assurances were enough for me. I promised that I would write to Stockmann and that I would stay in close touch with Van Beck. With that I took the next plane out of Bangkok and headed back to Chicago.

The letter was written and mailed only days later. It read as follows:

Mr. Hardy Stockmann
c/o Pat's Tavern
Near Chiang Mai Hotel
Chiang Mai, Thailand
Dear Mr. Stockmann,

Steve Van Beck has introduced us and I understand you need certain assurances before you will consider making the necessary introductions and passing on a message to your Chinese general friends somewhere in the Golden Triangle.

The Chicago-based Greek Heritage Foundation, of which I am chairman, is as you know engaged in a search for the long-missing Peking man fossils.

The Foundation is a cultural and educational organization, and it is nonpolitical. I was in Peking shortly after President Nixon when I first became interested in the missing fossils, but I am not

acting as an agent for the Peking government or any oth
government, nor am I a member of an investigative or intelligen
organization. I am a director of the United States People for t
United Nations which is interested in drug control, but I am in
way acting on their behalf in this instance.

Finally, what I have to say and the offer I am making is in
way connected with the United States government. I can assu
you that they would much prefer that I not establish any kind
communication with the Chinese Nationalists in the Gold
Triangle area.

The proposition that we are offering is this: $150,000 ca
reward on a no-questions-asked basis for the location of t
Peking man fossils, subject, of course, to complete authenticatio
and an agreement that the fossils, wherever their location may b
will be available to scientists all over the world for study.

If we can have some reasonable assurance that your contac
have these precious fossils, I will return immediately to Thailar
for discussions, and if the fossils are found, the reward can b
made immediately available. In addition to the $150,000 rewar
we will pay you $5,000 if the introductions you make result in t
finding and authentication of the fossils.

You can either communicate with me directly or through Ste
Van Beck at the *Bangkok Post*. I will be looking forward
hearing from you in any event.

With all good wishes,
Christopher G. Janus

Stockmann's reply followed shortly; for me, it was
mixed blessing, another twist which sent my search in y
another direction.

Mr. Christopher G. Janus
The Greek Heritage Foundation
Dear Mr. Janus,

I apologize for not having answered your letter earlier. Had
few assignments that were rather urgent, and only got back a fe
days ago.

I don't know what Steve Van Beck told you, but I think his enthusiasm seems to be carrying him away. I need neither assurances nor persuasions in order to help or follow up something interesting. I am a journalist and I might say a conscientious one. If Steve has some cloak and dagger ideas, they are entirely his own. (CIA, etc. I'm neither pro or anti and couldn't care less.)

I must disappoint you in that I have no Chinese general friends in the Golden Triangle. Of the five Kuomintang generals up here, I know two personally, but not too well. I do however have some sort of contact with some of the colonels in each general's group, and am usually well informed of what's happening.

I have already been to see Gen. Yen Yuan-tee, who retired some years ago and is now a wealthy businessman, very well informed about all Kuomintang movements. He laughed about my story and sincerely thinks that the fossils are not here. Gen. Lao Lee also has his residence here in Chiang Mai but hasn't been home for weeks. He prefers his camp up in the hills. The other three are permanently in the jungle and as you perhaps know, are fighting each other. Tuan is in the Maehonsong region, Lee (another one) northwest of Chiang Rai, and Na, the only one under direct control of Taipei, has his headquarters east of Chiang Rai.

I am in a position to visit them all but the trips are not easy. They involve travel by plane, jeep, mule, and hikes. Kuomintang passes have to be secured, guides hired and each trip would take between three and six days. Besides, the journeys would have to be kept from the public at least until completion, as Thai authorities do not like anybody roaming around in what they call "sensitive areas."

Should you want me to contact these generals, you would naturally have to fund these excursions. The direct cost involved is minimal but the loss of time for me is not.

I have also come across another possible lead, which you might find interesting. A friend of mine, a twenty-eight-year-old school teacher, told me of some bones she came across as a child. She was born near Kengtung in northeastern Burma, where her father ran a huge trading business having caravans going into Thailand, Laos, and China. Already as a young girl she was instructed in first aid and accompanied these caravans, officially as a nurse but really to spy for her dad on caravan leaders whom he apparently didn't trust. On one of her trips she says she had seen bones that were guarded by men with rifles and inspected by three doctors who had been brought from afar, one of them a Filipino. Her story was very vague and happened "long ago."

Since I am an amateur hypnotherapist and the young lady had for some time been one of my guinea pigs, I hypnotized her and regressed her to the time of the "bones." I taped the session with a few witnesses and she answered all my questions, asked in a Thai language I didn't understand. It was Nieu, a dialect spoken in Northwest Burma, which she then retranslated. Apparently her caravan had camped for the night next to another caravan on the banks of the Mekong River in the vicinity of Chiang Saen (40 miles northeast of Chiang Rai). Just before sunset a heavily guarded truck arrived and three blindfolded men emerged and were taken to a tent. As they were friendly with the other group they went over to look. There was a heap of bones in the tent and she understood that the three men were doctors that had been brought there to give "some sort of opinion" on the bones. She did not understand the language of the three visitors but got from the conversation that they were supposed to say whether or not the bones were human and what age they were. She also understood that the bones came from very far and were very precious.

She was eight years old at the time and didn't know what it was all about. She also didn't know where the bones were taken after that. Rather interesting. Unfortunately her father died in 1962, but she says that she still knows a few of her father's old hands, and on her next visit to her mother, who lives in Chiangkhong (on the Lao border) she will contact them and try to find out more.

That's all I can tell you so far. If something interesting comes up, I will inform you. Should you be interested in firsthand interviews with the generals, let me know.

With best regards,
Hardy Stockmann

As I read Stockmann's letter I was gripped with the same

puzzlement with which I had reacted to the hosts of other tales I had come across in my investigation. There seemed to be no end to such stories; the mystery seemed to have no bounds.

The school teacher's story would have to take a backseat to the generals. I was willing to back Stockmann for at least one trip to the area. The lead was too strong to go uninvestigated.

I contacted Steve Van Beck and asked him to relay my go-ahead to Stockmann. It was now up to these two men to find out what might lie in the Golden Triangle.

CHAPTER **15**
The Mystery

My search had now taken me to China, Taiwan, and Thailand in a little more than a year's time. Yet upon my return, I was not sure if I was any closer to Peking man than when I was first confronted with its loss in Choukoutien. I was certain only that I had closed some doors. I was now positive that neither the Chinese nor Taiwanese government knew where the fossils were, but that they each wanted them. There was no mistaking the sense of pride and cultural identity attached to the specimens. I had been thoroughly convinced that I would be a national hero if I recovered the bones, but now I was sure I could be a hero in Peking or Taipei.

There was little doubt in my mind after visiting the Nationalist Chinese museums that the fossils were not there. The atmosphere was all wrong, the individual personalities too concerned and outgoing to be involved in cover-up. And to what advantage, I asked myself. I could imagine the elaborate showcase they'd build for Peking man if he were in Taiwan. The Nationalist Chinese would flaunt the fossils as they did the art and cultural treasures they had taken from the mainland. They would use the fossils as much in defense of their present political ideology as the Maoists would in defense of theirs.

I was also convinced of the sincerity of the various Nationalist Chinese officials. George Yeh was eager to help in the search and gave me an entree to the country's museum treasures. The various museum directors were cordial and honored that I had come to talk with them. If their gestures of gratitude and their efforts at gaining the widest possible press coverage were part of an elaborate cover-up, then I was completely taken in by it. Even the aged Dr. Li had been eager to assist, although his thirty-year hunt for the fossils had apparently ground to a halt. And finally, Dr. Ho had revealed his story of the fossils and the Kuomintang general in Hong Kong. A rumor perhaps, but one which the museum director believed. I had little reason to doubt his sincerity.

The fact that I believed the Chinese Nationalists did not have the fossils in any official capacity, however, did not rule out suspicions that the fossils could nonetheless be found in Taiwan. They might be in the possession of a private citizen, one sworn to secrecy about their existence or, perhaps, so frightened that he had kept a thirty-year silence. The key to finding this person or persons—if they indeed did exist—was through informants not living in Taiwan. These informants, I speculated, might have already contacted me in some way. It was now up to me to ferret out the crucial details.

A number of people connected with the search believed that the fossils still were, in some form, in China. The logic behind such beliefs was sound: that the fossils were never taken from Camp Holcomb in the first place and were left to be scavenged by native Chinese; or if the Japanese did discover them, that they sold them to native Chinese. It was conceivable that the fossils had then been ground up into medicines as the fabled dragon bones had been, but it was equally likely that they had been sold or stolen. A number of laboratory specimens which had been lost turned up from time to time in thieves' markets and were ultimately returned to their original collections.

It was also quite possible, however, that if the fossils had remained in the Camp Holcomb area, they could have landed in the hands of Chinese who later evacuated China for Taiwan. They had not found their way into the official collections but could still be in private hands. There were, of course, untold variations to this theme, including the possibility that a person with the power of a Kuomintang general had gained possession of the fossils and sold or traded them. That thread led to the Golden Triangle. Such transactions could also have included Americans. Once the fossils were pronounced valuable booty, there was no end to the lengths various people would go to get possession of the priceless relics.

Upon my return from Taiwan and Hong Kong, I was convinced that the key to finding the fossils in China or Taiwan depended on individuals with specific information. No random search would uncover them; I had demonstrated that. I had received the widest possible publicity, the rumor mongers had come forward; now I had to reach those who really knew something.

The only place I had not gone to investigate significant leads was Japan. It was a logical site for such treasure, especially since Japanese anthropologists had shown marked interest in the fossils before the war broke out. Frank

Whitmore's discovery of fossil remains in Tokyo museums shortly after the war added to suspicions that Peking man had been taken by Japanese government officials. Yet the concentrated search by the Kempeitai and the harassment of Claire Taschdjian and other PUMC officials in 1942 and 1943 raised serious questions about whether or not the Japanese really knew what had become of Peking man.

My chief source of information about the Japanese role in the mystery was Howard Petersen, a managing editor for *Stars and Stripes*, who was stationed in Tokyo. Upon learning of the hunt, Petersen became hooked on the mystery and did some digging of his own in Japan. He uncovered the details of Lt. Albert Scalcione's postwar investigation, and through a series of letters to Claire Taschdjian, determined the exact details of the packing and shipment of the fossils to the marine compound in late 1941. Petersen freely shared his information with me and promised to dig even further. He was certain that if he found the Japanese company responsible for the takeover of Camp Holcomb on December 8, he might be able to find a soldier who could tell him something. But Japanese records were incomplete and Petersen had no success. He could not locate a single soldier who had participated in the takeover. He could not even find the names of the officers in charge.

Though Petersen's exhaustive research proved disappointing, I never closed the door on Japan. Dr. Li in Taiwan (who had mentioned his travels in Tokyo in search of clues) remained convinced that the fossils were there or had been at one time. Yet until more concrete information came up, I had no choice but to put Japan and the Japanese in the background and concentrate on the more solid leads I already had.

There were other stories which sounded equally credible. One was that the fossils had made their way to Manila, as suggested by Maria Sze, the young woman I had met in Peking. But this too was only a possibility, and one with no specific details to follow up. The most recent tale from Hardy Stockmann, related by his hypnotized friend, was fascinating, but also one with which I could do very little. The story was of a type that had surfaced with regularity during the hunt: Though supplied by reputable people, these tales were usually so vague and so remote that they could be used only as topics of conversation.

Another provocative story appeared in New York City in August 1973. I was contacted by a Russian-American man who identified himself as Alexis Petrov. He immediately impressed me as someone totally unlike the characters I had come to know in the search, for Petrov was straightforward and humorous and eager to help. He was a man in his early sixties, rather short at about five-foot-five, slightly chubby with a fair complexion. He said he was a teacher in San Francisco but that his family had been in the international import business, a profession I knew well, and that he had come across the Peking man mystery in a magazine article he read on a plane.

Petrov went on to explain that the article brought back vivid memories of the time he had taught biology at the Shanghai American School during 1936–37. For years, Petrov said, his family lived in Shanghai and engaged in trade between China and Russia. Most of the trade went from Shanghai to Sevastopol, Yalta, and the various Black Sea ports. Petrov himself was not involved in the business. Being a biology teacher, however, he was quite interested and informed about the Peking man discoveries in Choukoutien. He later heard about the North China marines and their responsibility for the fossils, and then that the Americans had been taken captive and placed in prison compounds in Tientsin and near Shanghai. Petrov said it was common knowledge that the Americans had been entrusted with the fossils, but he was not sure if the story had been passed on to his family along the waterfront or if he had learned it from faculty friends still at the Shanghai American School.

It was five years later, however, in 1946, that the most startling story circulated around Shanghai. Petrov said he heard that the fossils had been loaded onto a Russian ship docked in the Shanghai port and transported to the Soviet Union, perhaps to Yalta. He could not remember exact details—whether they had been smuggled aboard or if the Russians had somehow bargained for them—but he was certain of the basic facts.

Because of Petrov's background and his knowledge of the fossils, and because Shanghai's proximity to the prison camp at Kiangwan made it a strategic city in the mystery, I gave some credence to Petrov's story as another of the seemingly endless possibilities. One thing about the man stood out: At no time did he express fear, nor did he suggest that sinister occurrences were connected with the fossils' disappearance in Shanghai. He seemed forthright throughout the conversation and at no time gave me the impression that he was hiding anything. He also asked if the reward was still in effect. When he was told it was, he was satisfied and said he would renew the search in the Soviet Union, particularly in Yalta.

Petrov was off with a handshake, and I have not heard from him since.

Although I traveled across the world in my search, the most vital clues seemed to surface at my doorstep. That Andrew Sze, the Empire State Building woman, Dr. Foley, Herman Davis, and Claire Taschdjian should all be in New York City was remarkable. These persons seemed to have the most specific information, yet several of them were most difficult to deal with.

Sze had come forward in his own sincere yet ambivalent way, only to retreat and fade from the picture. His story will have to wait in the wings until he is ready or able to revive it.

It was a similar matter with Dr. Foley. He apparently preferred to let vital questions go unanswered. There seemed to be no way I could get any more information from him. I was told by State Department officials that if Foley would not budge in revealing the names of his friends in China to whom he had entrusted his marine baggage—if he would not even reveal the names to the FBI—there was little anyone else could do. Perhaps Foley has a point. If any Chinese had held onto the fossils through the years, even though they knew a search was on, they could now be in some kind of danger. It was impossible to second-guess Foley on this. Like Andrew Sze, Dr. William Foley kept the answer to himself.

I decided that only time could persuade Foley to tell all. If and when I was able to take another group to the People's Republic, I would ask him to accompany me. Every attempt would be made to meet his requirements for such a trip. If he were able once again to set foot in China, perhaps he could reestablish contact with the people he had dealt with before the war. Until then Foley would remain one of the most mystifying elements of the entire Peking man story.

The Empire State Building woman and her lawyer represented yet another possibility. The most recent statements of Dr. W.W. Howells and Professor Phillip Tobias about the authenticity of the skull shown in the woman's snapshot put added pressure on me to negotiate further with the elusive Harrison Seng. The most difficult of their demands, however, was the last one: that the government of the People's Republic assure them in writing that it would not attempt to prosecute the woman in any way for having the fossils. It was unlikely that such a letter would ever be written. In the meantime, Seng remained adamant about that requirement. Besides, he was now talking about $750,000 in payment for the footlocker. The prospect of raising it did not bother me as long as I could make sure that the woman's specimens would be positively authenticated before the transaction was made. But authentication hinged on the letter from the Chinese.

Harrison Seng continued to contact me from time to time, but there was nothing new to tell him as long as the Chinese remained silent. The entire matter had become a waiting game. The fossils had waited more than thirty years to be rediscovered, and in Harrison Seng's opinion, they could wait still longer.

Sifting through the various leads, the multitude of characters, the places, the rumors, the theories, and finally my own intuitions about the fate of the fossils, I could not help but construct my own scenarios for what may have happened. Each one differed, some only in a small detail, some with entirely different casts of characters, but each was based on some shred of evidence that had been verified along the way. It would have been easy, perhaps, to agree with Claire Taschdjian or Teilhard de Chardin and put the fossils squarely at the bottom of the Yellow Sea, the victims of careless cargo handlers. It would have been just as easy to assume that the fossils were lost in the confusion of the takeover of Camp Holcomb and were scattered in the dust. These assumptions supplied an uncomplicated, early death for Peking man, one that I had felt from the beginning was too convenient to accept.

I could have accepted more easily the theory that the Japanese took them or that they were sold or given to Chinese civilians in the Chinwangtao area. That theory dictated the slow, tedious tracking of people and places. There were any number of variations on this story, with the fossils ending up in China, Taiwan, Hong Kong, Thailand, Manila, the Soviet Union, or the United States. Trying to find out exactly where could deteriorate into a wild-goose chase for clues based largely on rumors, some true, others half true, some deliberately misleading or false. I was convinced that if the fossils had ended up with someone who did not want to produce them, there were hundreds of ways to keep them hidden.

The more logical explanation was the possibility that whoever had the fossils was keenly aware of his treasure, and that he would risk great danger if he revealed his possession. Any such person could have been thrown into the mystery by accident, through a friend or relative, and have no way to escape the consequences. This was probably the case if the fossils had been given to a Chinese citizen who feared punishment from his government. Because the People's Republic had been virtually sealed off from the West for thirty years, there would have been little opportunity for such an individual to dispose of the fossils outside the country.

If the person were a Nationalist Chinese, the reasons for keeping the fossils underground were more obscure. Perhaps fear, if the fossils had been taken illegally, or maybe a desire to collect some kind of reward. In such a case, the holder of the fossils would have to keep them hidden until the time was opportune; during an easing of the political tension in Taiwan and in the People's Republic, or when the proper people came forward with a sufficiently high ransom offer.

In all of these cases the basic premise is that the fossils do indeed exist. I feel sure that they do. I'm certain that human passions—whether patriotism or greed—have protected the fossils from being destroyed or thrown away. If I were to get to reach that hypothetical person, I knew from the start I'd have to appeal to the same motivations. If greed had been behind the disappearance of the fossils thirty years ago, greed would bring them out today. My reward offer was genuine and it alone had served to pull various clues out of the woodwork. I believed that was how I had found the Empire State Building woman even though, if her story was to be believed, it had been her husband and not she who had gone to great lengths to get the fossils.

Assuming the rumor concerning the Golden Triangle general could be satisfactorily checked and disproved, the evidence I had found closest to home seemed best. The personalities of Foley and Sze were complex and challenging, and the very appearance on the scene of the Empire State Building woman was remarkable. The answer to the mystery, I believed, might well be contained in this country.

Yet at a certain point in the search, another possibility came to mind. It was one which nagged at me, which grew to dominate my thoughts, as if I were feeding it with every new detail. There was a certain unity that struck me throughout the investigation: The various leads and individuals seemed somehow connected to each other, regardless of how disparate their lives appeared to be or how differently they fit into the mystery. I could not resist the temptation to look at the entire mystery in terms of one person, a person of immense intelligence and cunning, a person who could manipulate people and control situations. Such a mastermind could have devised the entire script in the thirty years that the fossils had been lost and then waited for the opportune moment to put it into action; such a mind would have immense patience and a supreme sense of timing and execution. He would be able to drop clues, scatter leads, and control agents who could set up scenes and situations that fit into the total order that he had devised.

The mastermind would have to have the most compelling motives for engineering such a scheme. They would have to include more than just money, certainly more than a playful sense of chicanery. He would have to have some kind of emotional relationship to the fossils, perhaps some drive for vengeance, as well as a measure of self-pity. The fossils had somehow taken a part of his life so dear and so irreplaceable that reparations would have to be made. If they were not, such a person would sooner see the fossils languish—or even destroyed—rather than let the world celebrate their re-emergence.

Such a mastermind would then be able to use agents such as the Empire State Building woman, for example. He would be able to turn the path of the investigation in a number of different directions to suit his purposes. There was no end to the levels, no end to the possible facets of this person's imagination, especially if each tactic, be it diversionary or legitimate, served to achieve prescribed results. Of course, my scenario changed depending on the various motives of the mastermind—money, revenge, power.

It would follow then that the entire mystery would be complex and frustrating at every turn for someone unaware of the existence of the mastermind and his master plan. Only when the maze is seen in its entirety does the mouse have a chance. I could have been the catalyst that started things in motion. I could have brought about the change in the fossils' status that the mastermind had been waiting for. Or I could be a pawn in a complex game with no ending; my efforts would bring Peking man no closer to and no further from the eyes of the world.

Perhaps such a theory was the product of a mind conditioned by television and cinema. On the other hand, the mastermind theory was so bizarre, so preposterous, that it might be accurate. But compelling as it was, the mastermind theory would have to be shelved as little more than an intriguing possibility.

Through the course of the investigation I have consulted with government agents, politicians, statesmen, scientists, investigators, reporters, and anyone else who could in some way contribute to the search. I have even sought the help of seers and psychics. My sources of information have been exotic and mundane. The trail has led me as far away as Bangkok and as near as the Empire State Building. I have delved into the distant past and into the recesses of my own memory. And if I should stumble across Peking man in my own attic it will be no less a find than if the elusive treasure emerges in the jungles of Thailand.

I have not abandoned hope that the fossils will be recovered or, failing that, be proven to be lost forever. Perhaps this book will play a part in the solution; perhaps the complete story of the search will be the lever that finally pries out the truth.

I realized the essence of Peking man when I first looked into the eyes of Dr. Wu, the director of the Peking Man Museum in Choukoutien. I could sense the significance of the fossils then, even though I knew nothing of their history or of their scientific importance. And throughout my search, I could feel the pulse of Peking man. At times it revealed itself in the zeal of an anthropologist poring over a laboratory cast, at times in the overwhelming fear in the face of an individual somehow involved in the mystery. I have experienced a keen excitement whenever a seemingly vital link surfaced and the fossils appeared to be close at hand. There has always been something very real about the fossils for me, something very alive.

The Chinese never do anything without a reason, I had been told. And they asked my help in finding the fossils.

I had been told the Chinese never throw anything away. I had traveled thousands of miles with that precept fixed in my mind.

One day I will meet Peking man. His bones will stare at me with unsparing lifelessness: so much yellowed, fossilized human remains. I will see in them the very energy and excitement that prevailed when Davidson Black and his fellows pried the first specimens from the limestone. I look forward to that day. Peking man has already slept too long.

Epilogue

Until the entire Peking man collection is accounted for, the search will go on. I have chosen to remain accessible, using any publicity that I might receive as a tool to generate further leads.

I recently received a call from a man who said that prior to 1949 he had been an aide to a Chinese general who showed him a footlocker full of fossils wrapped in blue silk. The man insisted they were the Peking man fossils and pressed me for the reward money.

I flew to Hawaii to meet him but he did not show up. He was, however, very insistent about the authenticity of the bones he had seen, and very definite about the blue silk wrappings. He was also very interested in the reward money, and I was suspicious of him because of that.

I have kept an eye on all previous leads. A trip to Thailand to meet with Hardy Stockmann has been arranged. I am finalizing plans to revisit the People's Republic, and have invited Professor Shapiro and Dr. Foley to accompany our group's search there.

The $150,000 reward for information leading to the recovery of the fossils still stands. I am committed to the continuing search for Peking man.

Always keep Ithaca fixed in your mind.
To arrive there is your ultimate goal.
But do not hurry the voyage at all.
It is better to let it last for long years;
and even to anchor at the isle when you are old,
rich with all that you have gained on the way,
not expecting that Ithaca will offer you riches.

Ithaca has given you the beautiful voyage.
Without her you would never have taken the road.
But she has nothing more to give you.

And if you find her poor, Ithaca has not defrauded you.
With the great wisdom you have gained, with so much
   experience,
you must surely have understood by then what Ithaca
   means.

—from "Ithaca" by Cavafy

Index